A TEXTBOOK OF

FISH AND FISHERIES

FROM BASIC ICHTHYOLOGY TO AQUACULTURE BIOTECHNOLOGY

FIRST EDITION

EDITORS

Dr. B. Singha
Head & Associate Professor
Department of Zoology
G. C. College, Silchar
ASSAM, INDIA

Dr. A. Dey
Head & Asstt. Prof. (Sel. Grade)
Department of Zoology
Karimganj College, Karimganj
ASSAM, INDIA

INVINCIBLE PUBLISHERS

First published in India in 2017 by Invincible Publishers

ISBN: 978-93-86148-61-2
Typeset in 11/14 Times New Roman

Publisher:
Invincible Publishers
G-120, Shushant Lok III
Sector 57, Gurgaon-122001
Website: http://www.i-publish.in/
E-mail: contact@i-publish.in
Phone: +91 124 4273677, +91 9599667779

Editors:
Dr. B. Singha
Dr. A. Dey

This book is dedicated to our all the beloved students

PREFACE

~~~ FIRST EDITION ~~~

All gill, paired and unpaired fins bearing aquatic craniate animals belonging to the superclass Pisces are collectively called fishes. Fishes are the world's first true and most common vertebrates comprising of more than 24,000 extant species. They began their evolution about 530 Ma during the Cambrian explosion. The jawed fishes showed maximum evolution and divergence in the Devonian period (419–359 Ma), and thus it is often regarded as the age of fish.

Human population is growing day by day and simultaneously aquatic habitats are also destroyed bit by bit. The production of enough protein-rich, nutritious food sustainably to feed everyone is one of the major challenges of the present day. Fishery sector offered a wide scope ranging from production of protein-rich food in the era of food insecurity to socioeconomic development. The concept of sustainable aquaculture (RAS, cage culture, pen culture, etc.) has evolved and grown in parallel with the growing evidence of overexploitation of wild fisheries. In the last 3 to 4 decades, spectacular growth took place in the aquaculture industry. The vibrancy of the industry in India can be reflected by the 11–fold increase in fish production in just six decades from 0.75 million tons to 9.6 million tons.

The present book offers insights into different aspects of the fishery industry and its sustainable development. The six chapters of this book cover the basic concepts like fish morphology, fish classification, the importance of the fishery industry, fish physiology, and migration, along with the reasons behind the depletion of fishery resources and laws and regulations in fisheries. The book also covers modern aquaculture techniques, aspects of biotechnology in fishery industry like stock improvement through androgenesis, gynogenesis, transgenesis, hybridization, polyploidization, etc. induced breeding of the IMC and the role of DNA barcoding in identification and traceability of seafood, monitoring the trafficking of threatened species by fish poachers and many more. The editors wholeheartedly thank all the authors of this book for their contribution and cooperation.

Silchar, Assam, India **EDITORS**

CONTENTS:

CHAPTER 1: FISH AND FISH CLASSIFICATION

CHAPTER 2: FISH MORPHOLOGY AND PHYSIOLOGY

CHAPTER 3: FISHERIES

CHAPTER 4: AQUACULTURE PRACTICES

CHAPTER 5: FISH DISEASES, FISH SPOILAGE AND FISH PRESERVATION

CHAPTER 6: AQUACULTURE BIOTECHNOLOGY

Chapter 1

FISH AND FISH CLASSIFICATION

Baby Singha[1,*], Anup Dey[2]

[1,*]Department of Zoology, G. C. College, Silchar, Assam, India
[2]Department of Zoology, Karimganj College, Karimganj, Assam, India

Chapter highlights:

All gill, paired and unpaired fins bearing aquatic craniate animals belonging to the superclass Pisces are collectively called fish. The term "finfish" is often used to distinguish the true fish from other aquatic life harvested in fishery industry such as molluscs, crustaceans, etc. or word ending with "fish" like shellfish, cuttlefish, starfish, jellyfish, etc. Fish are the world's first true and most common vertebrates comprising of more than 24,000 extant species. Fish are the first known true chordates and they have a long evolutionary history. They began their evolution about five hundred and thirty (530) million years (Ma) ago during the Cambrian explosion. The jawed fishes showed maximum evolution and divergence in the Devonian period (419–359 Ma) and thus it is often regarded as the age of fish. During the early Devonian period, jawed fishes had diverged into four distinct lineages: placoderms and spiny sharks, cartilaginous and bony fishes. Out of these clades, the former two are now extinct while the latter two that is cartilaginous and bony fishes which are still extant. Fishery industry is concerned with the catching, processing or selling of fish which has a significant economic importance, particularly as a protein-rich food source, scientific value, disease control, recreation, etc.

1.1 GENERAL DESCRIPTION OF FISH

1.1.1 Fish

All gill, paired and unpaired fins bearing aquatic craniate animals belonging to the **superclass Pisces** are collectively called fish. Most of the fishes are ectothermic (cold-blooded) that lives in water, breathes with gills, and usually has fins and scales. However, some fishes like white shark and tuna, which can maintain a higher core temperature (Goldman 1997; Carey and Lawson 1973).

- The term "**fish**" is generally used to describe multiple individuals of a single species. On the other hand term "**fishes**" is used to describe different fish species.
- The term "**finfish**" is often used to distinguish the **true fish** from other aquatic life harvested in fisheries like molluscs, crustaceans, etc. or word ending with "fish" like shellfish, cuttlefish, starfish, jellyfish, etc.

1.1.2 Evolution of fish

Fish are the world's first true and most common vertebrates comprising of more than 24,000 extant species. Fish are the first known true chordates and they have a long evolutionary history. They began their evolution about five hundred and thirty (530) million years (Ma) ago during the Cambrian explosion. The jawed fishes showed maximum evolution and divergence in the Devonian period (419–360 Ma), and thus it is often regarded as the age of fish. The evolutionary lineage of fish is briefly summarised below:

1) **Cambrian period (541–485 Ma):** The three genera *Pikaia*, *Haikouichthys* and *Myllokunmingia* which were closely related to fish were considered as the first ancestors of fish. These jawless groups had the basic vertebrate body plan with a notochord, rudimentary vertebrae, and a well-defined head and tail.

   These are followed by the appearance of Conodonts (cone-teeth, resembled primitive eels), about 495 Ma and Ostracoderms (shell-skinned, armoured jawless fishes) about 510 Ma. The Conodonts became extinct ~200 Ma, while Ostracoderms became extinct about 377 Mya (million years ago).

2) **Ordovician period (485–443 Ma):** Many extinct genus and class of jawless fish, including *Arandaspis*, *Astraspis*, Thelodonts and Pteraspidomorphi have appeared during this period. These jawless fishes had distinctive scales instead of armoured bodies, hypocercal tails, lateral line system and long and streamlined body.
3) **Silurian period (443–419 Ma):** Many armoured and nonarmored **jawless** fishes appeared during this period including extinct classes like Anaspida (lacked heavy bony shield, paired fins, but had hypocercal tail), Osteostraci (bony shields and paired fins), spiny sharks, etc. In addition to this, armoured jawed fishes of the class Placodermi (plate-like skin) also appeared at the end of the Silurian period. However, they became extinct at the end of the Devonian period.

   The earliest known bony fish *Guiyu oneiros* and the earliest known ray finned fish (Class- Actinopterygii) *Andreolepis hedei* also appeared in the late Silurian period.
4) **Early Devonian period (419–393 Ma):** The first appearance of many extinct lobe-finned fish (Class- Sarcopterygii) like *Psarolepis*, *Holoptychius*, *Laccognathus,* etc. occurred at the start of the Devonian period. Fishes belonging to the two orders Ptyctodontida (order of unarmored placoderms) and Petalichthyida (order of flattened placoderms) also appeared in the early Devonian, but they became extinct at the end of the Devonian period.
5) **Middle Devonian period (393–383 Ma):** During the middle of the Devonian period, both marine and freshwater jawed fish were increasing in diversity. Cartilaginous fishes like sharks, rays and chimaeras (Class- Chondrichthyes) appeared during this period. On the other hand, the diversity of armoured jawless Ostracoderms fishes declined during this period.
6) **Late Devonian period (375–360 Ma):** Due to a prolonged series of extinctions during the late Devonian period, it had a significant impact on vertebrate faunas including fishes. It is estimated that approximately 75% of all fish families disappeared during due to late Devonian extinction. As a result over 70% of fish species living in the Devonian were no longer existed in the Carboniferous Period.

   The late Devonian extinction events primarily affected the marine community. For example, the genus of marine Placoderms like *Dunkleosteus*, *Titanichthys*, *Materpiscis,* etc. became extinct in the Late Devonian. In terms of extant fish species, Chondrichthyes (cartilaginous fish) make up only around 1050 species, while Osteichthyes (bony fish) constitutes around 28,000 species. Thus, the bulk of the extant fish is from the class Osteichthyes, which includes everything from tuna to eels. In the evolutionary time scale, these fish evolved around more that 410 Mya (Late Silurian), but attained diversity in the middle Devonian period when Placoderms and huge sharks began to lose their dominance (Brough 1936; Long John 1996).

Besides diversity, all bony fishes share a hydrostatic organ called swim bladder. From this point; the bony fish diverged into three lineages.

1) First group: It includes the most of the common ray-finned fishes such as tuna, bass, salmon found in both fresh and salt water.
2) Second group: It includes the lungfish found in freshwater.
3) Third group: The last group includes the lobe-finned fish.

1.1.3 Economic importance of fish

Fishery industry is concerned with the catching, processing or selling of fish. The economic importances of fish are:

1) **As a protein rich food:** Fish are an important source of protein, minerals, and vitamins A and D for humans worldwide.
2) **Ornamental fishery:** Ornamental fish keeping is one of the most popular hobbies in the world today. In this regard, ornamental fishes form an important commercial component of aquaculture for aesthetic requirements.
3) **Sports fishing or recreational fishing:** The traditional sports fishing is enjoyed by millions of people worldwide. Fishing primarily for pleasure or competition is called recreational fishing, also called sports fishing.
4) **Medical importance:** Watching ornamental fish aquarium has a number of health benefits like it reduce stress, lowers blood pressure, reduces hyperactivity disorder in children, Alzheimer's disease, etc. The aquarium is widely used in "Pet therapy" to reduce loneliness, anger, depression, and stress.

5) **Scientific value:** Fish like lungfish has zoological importance because of their discontinuous distribution and unique anatomical features.
6) **Disease control:** The fish species that feed on the larva of insects such as mosquito are called larvivorous fish. Larvivorous fish like *Gambusia*, *Trichogaster*, *Channa punctatus*, *Colisa fasciatus*, *Puntius ticto*, *Chanda nama,* etc. plays a significant role in controlling the mosquito vector of diseases like malaria, yellow fever, etc.
7) **Fishery byproducts:** Fish industry also yields a number of byproducts like body oil, liver oil, fish emulsion, fish glue, isinglass, fancy articles and many more for commercial purposes.

1.2 FISH CLASSIFICATION

1.2.1 Classification of animal kingdom

The phylum Chordates includes animals that have a bilateral body plan and possess at some stage in their lives all of the following characters:

1) **Notochord:** It is the main axial support of the body. It is a stiff cartilaginous rod which extends along the inside of the body. In subphylum Vertebrata, the notochord is replaced by a bony structure called vertebral column or backbone or spine. It is composed of 31 segmented series of bones called vertebrae.
2) **Nerve cord:** It is a hollow fluid filled cord lying dorsal to the notochord. In fish and other vertebrates, nerve cord develops into the spinal cord.
3) **Pharyngeal slits:** These are openings lying between the pharynx and the outside. In fish, the slits are modified into gills, but in invertebrate chordates (e.g., lancet, tunicates) they are part of a filter-feeding system.
4) **Endostyle:** It is a longitudinal ciliated groove lies on the ventral wall of the pharynx. In filter-feeding species, it produces mucus to gather food particles. It also stores iodine, and thus it may be a precursor of the vertebrate **thyroid gland**.
5) **Post-anal tail:** A muscular tail called post-anal tail extends in the backwards direction behind the anus.

The phylum Chordata is classified into two major groups: A) Acraniata or Protochordata and B) Craniata or Euchordata.

A] GROUP- ACRANIATA/ PROTOCHORDATA: The animals belonging to group Acraniata has the following unique features:

a) All are marine, small, primitive and lower chordates.
b) They don't have skull/ cranium.
c) They Lack vertebral column, jaws, and brain.
d) They lack appendages.

The group **Acraniata** is divided into **two (2) Subphylum** 1] Urochordata or Tunicata and 2] Cephalochordata

1] Subphylum- Urochordata or Tunicata [commonly known as sea squirts]:

a) Notochord present only in the larval stage.
b) Body of the adult is covered by a cuticular tunic or test.
c) Presence of numerous gill slits.
d) Coelom is completely absent.

The subphylum **Urochordata** is subdivided into **three (3) classes** a) Larvacea b) Ascidiacea c) Thaliacea

1) Class 1- Larvacea:

a) Free swimming pelagic form.
b) Test is temporary.
c) Single pair of gill slit is present.
d) Presence of tail

Example- Animal belonging to class Appendicularia (*Oikopleura*).

2) Class 2- Ascidiacea:

a) Fixed or free-swimming marine form.
b) Test is permanent and well developed.
c) Branchial sac has numerous gill slits.
d) Tail absent.

Example- *Ascidia*, *Herdmania*, *Pyrosoma*, etc.

3) Class 3- Thaliacea:

a) Free swimming pelagic form.
b) Test is permanent and transparent.
c) Branchial sac has either two large or numerous small gill slits.

d) Life history exhibits an alternation of generation.
Example- *Doliolum*, *Salpa*, etc.

2] Subphylum- Cephalochordata:

a) Notochord and nerve cord present throughout life along the entire length of the body.
b) Body is fish like and segmented.
c) Presence of numerous gill slits.

The subphylum **Cephalochordata** has a **single (1) class** called **Leptocardii.** It has the following unique features

a) Body fish like, segmented.
b) Free swimming and burrowing animal.
Example- *Branchiostoma* (lancelets), *Asymmetron*, etc.

B] GROUP- CRANIATA/ EUCHORDATA: The animal belonging to group Craniata has the following unique features:

1) They are aquatic or terrestrial large sized higher chordates.
2) They have cranium or skull.
3) Usually, possess 2 pairs of appendages.

The group **Craniata** includes a single **subphylum Vertebrata.** It has the following unique features

1) The notochord is replaced by the vertebral column.
2) Body divisible into head, neck, trunk and tail.
3) Usually dioecious.

The subphylum **Vertebrata** of group Craniata is divided into **two (2) divisions** a] Agnatha and b] Gnathostomata

a] Division- Agnatha:

1) Vertebrates without jaws.
2) Paired appendages are absent.
3) Inner ear with 2 semicircular canals.

The division Agnatha of subphylum **Vertebrata** is subdivided into **two (2) classes** 1) Ostracodermii and 2) Cyclostomata

1) **Class 1- Ostracodermii:**

a) They have a fish-like body with heavy head armour.
b) Presence of bony dermal plates (large scales) in the skin.
c) Two paired eyes and a median pineal eye was present.
Example- *Cephalaspis, Pituriaspis, Neeyambaspis*, etc.

2) Class 2- Cyclostomata:

a) Body is long, rounded and eel-like.
b) Skin without any scales
c) Mouth is suctorial type with horny teeth.
Example- *Petromyzon* (sea lamprey), *Myxine* (hagfish), etc.

b] Division: Gnathostomata:

1) Vertebrates with jaws
2) Presence of paired appendages
3) Inner ear with 3 semicircular canals.

The division **Gnathostomata** is subdivided into **two (2) Super-classes** 1) Pisces and 2) Tetrapoda

1) Superclass- Pisces:

a) Body is streamlined and possesses both paired (dorsal, caudal and anal) and unpaired (pectoral and pelvic) fins.
b) The skin is covered with ctenoid or cycloid or placoid scales.
c) Presence of gill slits (usually five pairs).
d) Presence of well developed lateral line sense organ.
e) A swim bladder or air bladder is usually present with some exceptions.

The superclass **Pisces** is subdivided into **three (3) classes**: a) Placodermi [Extinct], b) Osteichthyes and c) Chondrichthyes.

2) Superclass- Tetrapoda:

a) Presence of pentadactyle limbs.
b) Presence of cornified skin.
c) Paired and unpaired fins absent

The superclass **Tetrapoda** is subdivided into **four (2) classes**: a) Amphibia, b) Reptilia, c) Aves and d) Mammalia

1.2.2 Classification of Superclass Pisces

The animal belonging to superclass Pisces are the representative of fresh waters, marine waters and brackish waters aquatic vertebrates. Characteristics of superclass Pisces are given below:

1) Body is streamlined and possesses both paired (dorsal, caudal and anal) and unpaired (pectoral and pelvic) fins.
2) The skin is covered with dermal scales like ctenoid or cycloid or placoid scales.
3) The body is streamlined, but some are elongated snakelike and a few are dorsoventrally compressed.
4) The endoskeleton is bony or cartilaginous.
5) Presence of gill slits usually five pairs.
6) Presence of well developed lateral line sense organ. It helps them to know the disturbances in the surrounding aquatic environment.
7) A swim bladder or air bladder is usually present, but secondarily lost in some species.
8) Heart is venous and two chambered: 1 auricle and 1 ventricle. Thus, single circulation of blood occurs.
9) Sexes are separate. Fertilisation is either external or internal.
10) Presence of gonads and true gonoducts.
11) They respire mainly by gills. In addition, some fishes that occasionally stay out of the water possess system of air chambers called accessory respiratory organs. These are formed by the outgrowths from the mouth or gill region.

    Different types of accessory respiratory organ found in fish are- buccopharyngeal epithelium (e.g., mud skippers) gut epithelium (e.g., *M. fossilis*, *L. guntea*) and brachial diverticulum like labyrinthinie organ, arborescent organs (e.g., *Anabas testudineus*, *Clarias batrachus*, etc.)
12) Presence of ten pairs of cranial nerves.

Different authors have provided different schemes of classification. Some divided it into 6 classes (Dipnoi, Teleostei, Ganoidae, Elasmobranchi, Marshipobranchi, and Leptocadii), while other classified superclass Pisces into 7 classes (Pterichthys, Coccostei, Acanthodii, Elasmobranchii, Holocephali, Dipnoi, and Telestomi). However, the most simple and widely accepted classification scheme includes only **three (3) classes**: a) Placodermi [Extinct], b) Osteichthyes and c) Chondrichthyes.

A] Class- Placodermi [Extinct]:

1) It includes all extinct armoured prehistoric fish.
2) Their head and thorax were covered by articulated armoured plates.

3) The rest of the body is either scaled or naked.
4) Placoderms were also the first fish to develop pelvic fins and true teeth.

Example- *Coccosteus*, *Fallacosteus*, *Rolfosteus*, *Materpiscis,* etc.

B] Class- Chondrichthyes (Cartilaginous fish):

1) Tail is heterocercal.
2) Skin with placoid scales.
3) Gill slits are not covered by an operculum.
4) Presence of pelvic clasper in male.
5) Endoskeleton is cartilaginous.

Example- *Scoliodon* (Dogfish), Shark, *Torpedo* (Electric ray), etc.

C] Class- Osteichthyes (Bony fish):

1) Tail is homocercal.
2) Skin is covered by cycloid or ctenoid scales.
3) Gill slits are covered by operculum.
4) Pelvic clasper in male is absent.
5) Endoskeleton is bony.

Example- *Labeo rohita, Catla catla, Anabas testudineus,* etc.

1.3 FISH CLASSIFICATION BASED ON FEEDING HABIT, HABITAT AND MANNER OF REPRODUCTION

1.3.1 Classification

Fish, the first vertebrates on Earth evolved around 500 million years ago. According to FishBase database, 33,400 species of fish had been described by January 2016. Thus, fish is the most diverse group of vertebrates in terms of species. It has been estimated that approximately 80% of the fish population of the globe is represented by the Indian fishes. However, fish species differ from each other in terms of their feeding habit, habitat and manner of reproduction.

1.3.2 Classification on the basis of feeding habit

The natural food of fish includes detritus, plankton, worms, insects, aquatic plants and smaller fish. However, this natural food of fish may be its basic food, occasional or secondary food, incidental food and obligatory or emergency food of a fish.

Fish species can be further categorised into three types on the basis of the extent of variety in their food (Nikolsky 1963).

1) **Monophagic fish:** These fishes feed on a single type of food.
2) **Euryphagic or polyphagic fish:** These fishes feed on a wide variety of food items.
3) **Stenophagic fish:** These fishes feed on a few selected types of food items.

On the basis of the type of food preference, fishes can be classified into the following five basic eating groups:

1) **Herbivorous fish:** They feed primarily on plant based food such as phytoplankton, filamentous algae and higher vascular plants (macrophytes).
   Example- Surgeonfishes, Silver carp, Grass carp, Herrings, etc.
2) **Carnivorous fish or Predatory fish:** They feed primarily on animal based food such as zooplankton, smaller fish, mosquito larva or fruit flies, earthworms, insects, etc.
   Example- catla, perch, muskie, pike, walleye, salmon, etc.
3) **Omnivorous fish:** They feed on both plant and animal based food. Most aquarium fish are omnivores and they eat both animal based food and vegetables. They can be further classified into 2 types:
   a) **Herbi-omnivorous:** They feed more on plant food than the animal based food items. Example- *Puntius sarana.*
   b) **Carni-omnivorous:** They feed more on animal food than the plant based food items. Example- *Oreochromis mossambicus.*
4) **Detritivorous fish:** Detrivores, detritus feeders, or detritus eaters feed primarily on detritus i.e., dead and decaying organic matter.
   Example- Common carp, Mrigal, etc.
5) **Limnivorous fish:** They feed on vegetable matter and micro-organisms found in mud. They are commonly known as mud-eaters or bio-film grazers.
   Example- Loaches, Catfish, Blennies, etc.

On the basis of specific type of food (plankton, insect, fish, etc.), fish can be more specifically classified into groups like

1) **Piscivorous fish:** Fish species which feed entirely on other small fishes are called piscivorous fish.
   Example- *Channa striatus*, Bull trout (*Salvelinus confluentus*), Flatfishes (order Pleuronectiformes), Needlefish (family Belonidae), etc.
2) **Larvivorous fish:** Fish species which feed entirely on insect larvae are called larvivorous fish.
   Example- Mosquitofish, *Cyprinus carpio*, *Oreochromis niloticus*, etc.
3) **Planktivorous fish:** Fish species which feed entirely on plankton (zooplankton or phytoplankton) are called planktivorous fish.
   Example- Catla, Silver carp, etc.
4) **Cannibalistic fish:** Fish species which consume individual of the same species as food are called cannibalistic fish.
   Example- *Clarias batrachus*, three-spined sticklebacks, tessellated darters, etc.
5) **Malacophagus fish:** Fish species which feed entirely on mollusca are called malacophagus fish.
   Example- *Tor tor.*
6) **Carcinophagus fish:** Fish species which feed entirely on crustaceans are called carcinophagus fish.
   Example- Black bass.
7) **Insectivorous fish:** Fish species which feed entirely on insects are called insectivorous fish.
   Example- Mosquitofish (*Gambusia affinis* and *G. holbrooki*), Killifish, Rainbowfish, Tetras, etc.

On the basis of mechanism of feeding they employed, fishes can be classified into following five kinds

1) **Predators:** They feed on macroscopic animals. They grasp and hold their prey firmly by their highly developed teeth.
   Example- Ribbon fish, Seer fish, Barracuda etc.
2) **Grazers:** In this type of feeding habit, fish take their food by bites on food items like coral polyps, algae, other fishes etc.
   Example- Butterfly fishes, Parrot fishes, Trout, Salmon, etc.
3) **Strainers:** The fish belong to this group possess well developed gill rackers by which they strain or sieve incoming water to obtain planktons.
   Example- Silver carp, Mackerel, etc.
4) **Food suckers:** They have well developed lips by which they suck

their desired food or food containing material.
Example- Sturgeon, Catfish, etc.

5) **Parasitism:** The parasitic fish possess a well developed mouth by which they remain attached to the host body and pierce and suck the body fluid from the host's body.
Example- Deep sea angler fish (*Ceratias*)

In addition to this, fishes can be further classified into following three kinds on the basis of trophic niche in the water body; they occupy (Das and Moitra 1955; 1956).

1) **Surface feeder:** They feed on food items available in the surface layer of the water body.
Example- Catla, Silver carp, etc.
2) **Mid-column feeder:** They feed on food items available in the mid-water or column water of the water body.
Example- Rohu, Grass carp, etc.
3) **Bottom feeder:** They feed on food items available in the bottom layer of the water body.
Example- Mrigal, Common carp, etc.

1.3.3 Classification on the basis of habitat

Fishes have their habitat in lakes, streams, oceans, and estuaries. According to habitat, fishes can be classified into the following classes:

1) **Freshwater fish:** These fishes are found in the water bodies such as ponds, lakes, hill streams, rivers, wetlands, etc. which has salinity less than 0.05%. More than 41.24% of all known species of fish are freshwater fish. The body fluid of freshwater fish is hypertonic to their external watery environment. The main physical adaptations found in freshwater fish are:
    - Scales on their body: to reduce water diffusion through their bodies.
    - Well-developed kidneys: to produce large amounts of dilute urine.
    - Special cells in gill lamellae: to uptake of ions like sodium and chloride from the water.

    Example- Rohu, Catla, Mrigal, Grass carp, etc.

2) **Marine fish:** These fishes are capable of living in water bodies such as sea water, which has salinity in the range of 3.1% to 3.8% and pH in the range of 7.5 to 8.4. The body fluid of freshwater fish is hypotonic to their external environment. The main physical adaptations found in freshwater fish are:
   - Kidney with relatively few small glomeruli: to produce less amounts of urine.
   - Secretory cells in the gills: to secrets of ions like sodium and chloride into water.
   - Rectal Gland/ salt gland: to excrete excess salts from body fluid.
   - Production and accumulation of metabolic urea, TMAO etc. in body fluid.

   Example- Tuna (*Thunnus atlanticus*), Mackerel (*Scomber colias*), Cod (*Gadus morhua*), Herring (*Clupea pallasii*), etc.
3) **Brackish water fish:** These fishes are capable of surviving in the brackish water bodies like estuaries, mangroves, brackish seas, and lakes, etc. which has a salinity in between 0.05 to 3%. In India, brackish water fish cultivation is carried out in **bheries** (man-made impoundments in coastal wetlands) of West Bengal and **pokkali** (salt resistant deepwater paddy fields along the Kerala coast).

   Example- Hilsa (*Tenualosa ilisha*), Asian seabass popularly known as Bhetki in India (*Lates calcarifer*), Milkfish (*Chanos chanos*), Grey mullet (*Mugil cephalus*), etc.
4) **Migrating fish:** Fish species migrating between marine and fresh waters for the purpose of breeding are called migrating fish. They can be either anadromous (migrate from the sea to fresh) or catadromous (migrate from fresh to the sea) or amphidromous (migrate from fresh water to the seas, or vice versa, but not for the purpose of breeding).

   Example: American eel, European eel, etc. (catadromous) and Atlantic salmon, Pacific salmon, etc. (anadromous).
5) **Tropical or warm water fish:** Fish species found in aquatic tropical environments around the world are called tropical fish. It includes both freshwater and saltwater fish species. Tropical fish are popularly selected for aquariums due to their bright coloration.

   Example- Angel, Killifish, Tetras, Swordtails, Goldfish, etc.
6) **Cold water fish:** These fish species prefers cooler water temperatures than tropical fish, typically below 20°C (68°F). They can even withstand temperature down to 10°C.

Example- Koi, Gobio, Goldfish, Gras carp, Three-spined stickleback, etc.

7) **Pelagic fish:** These fishes prefer to live in a zone of the ocean or lake waters that is neither close to the bottom nor near the shore called pelagic zone. They can be further classified into:

   a) **Epipelagic fish:** These fishes inhabit the epipelagic zone or photic zone (from the surface down to around 200 metres).
      Example- Mackerel, Tunas, Pomfrets, Herring, Salmon, Whale sharks, etc.

   b) **Mesopelagic fish:** These fishes inhabit the mesopelagic zone (from 200 m down to around 1,000 m).
      Example- Lanternfish, Ridgehead, Sabretooth, Stoplight loosejaw fish, etc.

   c) **Bathypelagic fish:** These fishes inhabit the bathypelagic zone (from 1,000 m down to around 4,000 m).
      Example- Bristlemouth, Anglerfish, Viperfish, Telescopefish, Gulper eel, etc.

8) **Demersal fish:** These fishes prefer to live on or near the bottom of seas or lakes called demersal zone. Their habitat usually consists of mud, sand, gravel or rocks sea floors and lake beds. They can be further classified into:

   a) **Benthopelagic fish:** They inhabit the water just above the bottom. They have neutral buoyancy and generally feed on benthos and zooplankton.
      Example- Rattails, Cusk eels, Deep sea cods, Deep sea eels, Halosaurs, Notacanths, etc.

   b) **Benthic fish (or groundfish):** They inhabit the lowest level of the water body, i.e., sea floor or lakebed.
      Example- Dragonets, Flatfish, Stingrays, Great hammerhead, Tripodfish, Flathead fish, etc.

1.3.4 Classification on the basis of manner of reproduction

Both external and internal modes of fertilization are found in fish. Many fish species have modified their fins to deliver spermatozoa into the female's cloaca during breeding season. For example, in male dogfish, the posterior portions of both pelvic fins are modified into an intromittent or copulatory organ called claspers. They used claspers to deliver semen into the female's cloaca during mating.

On the basis of different reproductive strategies found in various fish species, they can be classified into the following categories (Lode 2012):

1) **Ovuliparous fish:** Male and female individuals belonging to this group releases their gametes into the surrounding water for external fertilisation. Thus, fertilisation takes place outside the mother's body.
   Example- Salmon, Cichlids, Tuna, Eels, etc.
2) **Oviparous fish:** Around more than 97% of all known fish are oviparous in which fertilisation occurs internally. Males of these fishes possess a well developed intromittent organ for sperm delivery.
   Example- Claspers (modified pelvic fins) found in Dogfish.
3) **Ovoviviparous fish:** The fish species belongs this group retains their fertilised egg (embryo) inside the mother's body until they are ready to hatch. But unlike mammal, there is no placental connection with the mother and the embryo receive their nourishment from the yolk sac.
   Example- Guppies, Angel sharks, Coelacanths, etc.
4) **Viviparous fish:** In this group of fish, development occurs within the mother's body, and thus they give birth to live young ones. The viviparous fish can be further classified into two types:
   a) **Histotrophic viviparous:** In this group, during development inside the oviduct the embryo obtains nutrients by consuming other tissues like ova or zygotes.
      Example- Shortfin mako, Porbeagle, Halfbeak, etc.
   b) **Hemotrophic viviparous:** In this group, during development inside the oviduct the embryo obtains nutrients from parents by a structure similar to the placenta.
      Example- Surfperches, Lemon shark, Pipefish, etc.
5) **Hermaphroditic fish:** Many fish species possess both male and female reproductive organs and are called hermaphroditic fish. Hermaphroditism has been reported to occur in 14 families of teleost fishes (Shapiro 1984). Though the exact mechanism is not known, but sex reversal or sequential hermaphroditism is found in many fish species in natural condition. It can be of two types:
   - Protandry: Male to female sex reversal
   - Protogyny: Female to male sex reversal

For example, Goby, groupers, parrotfishes and wrasses fish population generally contains a number of females and a single dominant male. With the loss of the male from the group, one of the adult female fish undergoes sex reversal to become the male of the group. Similarly, water temperature also influences sex reversal in adult zebrafish.

6) **Sexually parasitic fish:** Deep sea ceratioid anglerfish exhibit a peculiar and unique mode of reproduction. Males of this fish are much smaller (6.2 mm) in size than the females (46 mm). When a male finds a female, the male attaches itself to the body of female typically in her belly region. Sometimes many males attach to the body of a single female.

   After attachment, fusion of epidermal tissues, and union of the circulatory systems of male and female occurs so that the male becomes completely dependent on the female for blood-transported nutriment. At the time of reproduction, both male and female releasing their sperm and eggs into the water at the same time, and thus fertilisation takes place externally (Pietsch 1975).

7) **Parthenogenetic fish:** Some fish reproduces parthenogenetically in which growth and development of embryos occur without fertilisation or development of young from unfertilised egg (Hubbs and Hubbs 1932).

   Example- Bonnethead shark, Zebra shark, Amazon molly. In addition to these, hammerhead and blacktip sharks also show facultative parthenogenesis.

1.4 REFERENCES

Brough J (1936) On the evolution of bony fishes during the triassic period. Biological reviews. 11(3): 385-405.

Carey FG, Lawson KD (1973) Temperature regulation in free-swimming bluefin tuna. Comparative Biochemistry and Physiology. 44(2): 375–392.

Das SM and Maitra SK (1955) Studies on the food of some common fishes of Uttar Pradesh, India (Part I). Proceedings of the National Academy of Sciences, India. 25: 1-6.

Das SM and Maitra SK (1956) Studies on the food of some common fishes of Uttar Pradesh, India (Part II). Proceedings of the National Academy of Sciences, India. 26: 213-223.

Goldman KJ (1997) Regulation of body temperature in the white shark, *Carcharodon carcharias*. Journal of Comparative Physiology. 167 (6): 423–429.

Hubbs CL and Hubbs LC (1932) Apparent parthenogenesis in nature, in a form of fish of hybrid origin. Science. 76 (1983): 628–630.

Lode T (2014) Oviparity or viviparity? That is the question...". Reproductive Biology. 12 (3): 259–264.

Long John A (1996) The rise of fishes: 500 million years of evolution, 5th edn. University of New South Wales Press Pty Limited, Sydney, Australia. Pp: 1-287.

Nikolsky GV (1963) The ecology of fishes. Academic Press, London, 1st edn. United Kingdom. pp: 1-352.

Pietsch TW (1975) Precocious sexual parasitism in the deep sea ceratioid anglerfish, *Cryptopsaras couesi* gill. Nature. 256: 38–40.

Shapiro DY (1984) Sex reversal and sociodemographics processes in coral reef fishes. In: Potts GW, Wootoon RK, eds. Fish reproduction: Strategies and tactics, Academic Press, pp: 103–116.

*** ***** ***

Chapter 2

FISH MORPHOLOGY AND PHYSIOLOGY

Nitu Debnath[1,*], **Joyobrato Nath**[1]

[1,*]Department of Zoology, Cachar College, Silchar, Assam, India

Chapter highlights:

Presence of paired and unpaired fins is the principal characteristic features of the superclass Pisces. Beside their principal functions like propelling, steering and turning the body in water during swimming, they are modified in different fishes to play the key role in gliding (flying fish), delivering sperm (sharks and mosquitofish), injecting venom (stonefish) and walking (anglerfish). Swim bladder found in fish mainly function as a hydrostatic organ, enabling the fish to maintain its water depth without sinking and expending much energy. Osmoregulation in Elasmobranch is usually achieved by accumulating organic nitrogenous compounds, such as urea and TMAO in their body fluid and by salt excretion and uptake by the rectal gland, kidney and gill. The generation of electricity in electric fishes is mainly utilised for electrolocation, self-defence, electrocommunication and prey stunning, while bioluminescence is used for camouflage, mimicry, defense, detection of prey, and to attract mates. Similarly, schooling behaviour offers numerous benefits to individual fish, including foraging advantage, reproductive advantage, and increased protection from predators. Fish parental care ranges from hiding of eggs, protection of fry, cleaning, fanning of eggs, carrying young to the feeding of young. Migration in fish enables fish to explore new favourable areas, increases genetic diversity, reduces predation pressure, etc.

2.1 FINS, THEIR TYPES AND MODIFICATIONS

2.1.1 Fins

The presence of paired and unpaired fins is the principal characteristic features of the superclass Pisces. Fins are paired or unpaired appendages protruding from the body of fishes. All the fins are supported only by muscles except the caudal fin which is directly connected to the spine. In bony fishes (Osteichthyes), the fins are either joined to the body by a single bone (Sarcopterygii) or they contain spines or rays (Actinopterygii).

Beside their principal functions like propelling, steering and turning the body in water during swimming, they are modified in different fishes to play the key role in gliding (flying fish), delivering sperm (sharks and mosquitofish), injecting venom (stonefish) and walking (anglerfish).

2.1.2 Types of fins in fishes

Fins of adult fishes can be broadly classified into two types- median unpaired fins and paired lateral fins.

A] Lateral paired fins: Fishes have two sets of paired fins and these include anterior pectoral fins and posterior thoracic or abdominal pelvic fins.

1) **Pectoral fins (paired):** Two pectoral fins are located just behind the operculum of each side. The pectoral fins are homologous to the forelimbs of tetrapods.

   Depending on the species, pectoral fins assist in maintaining depth (sharks), flight (flying fish; order Exocoetidae) and walking (anglerfish and mudskipper).

2) **Pelvic fins or Ventral fins (paired):** The paired pelvic fins are characteristically located ventrally behind the pectoral fin on each side. The pectoral fins are homologous to the hind limbs of tetrapods. This paired fin help in balancing the fish during swimming, going up or down through the water. If these pelvic fins are present in the thoracic region, just below the anterior pectoral fins they are called thoracic pelvic fins and if these are present in the abdominal region, just in front of anus they are called abdominal pelvic fins.

   In some fish, these fins are modified, reduced or even absent. For example, the reduced or absent pelvic fin in eel-like fish allows it to squeeze through small crevices.

B] Median unpaired fins: In the majority of fishes these include one, two, or three dorsal fins, a distinct caudal fin, an anal fin and a less common adipose fin.

1) **Dorsal fin:** It is located on the dorsal side of the body, generally between the pelvic and pectoral fins. The number of dorsal fins varies among species; many possess single dorsal fin, while some possess 2-3 dorsal fins (Example- order Acropomatidae). It is used in steering where it prevents fish from side to side rolling and assists in sudden turns during locomotion.
2) **Anal fin:** It is located ventrally behind the anus. During active swimming, this fin is used to stabilise the body (Fig. 2.1).
3) **Adipose fin:** It is a soft, fleshy fin found dorsally behind the dorsal fin and in front of caudal fin in Salmonidae, characins and catfishes.

   The presence of neural network in the fin indicates that it may have a sensory function in fishes (Buckland-Nicks et al. 2011).

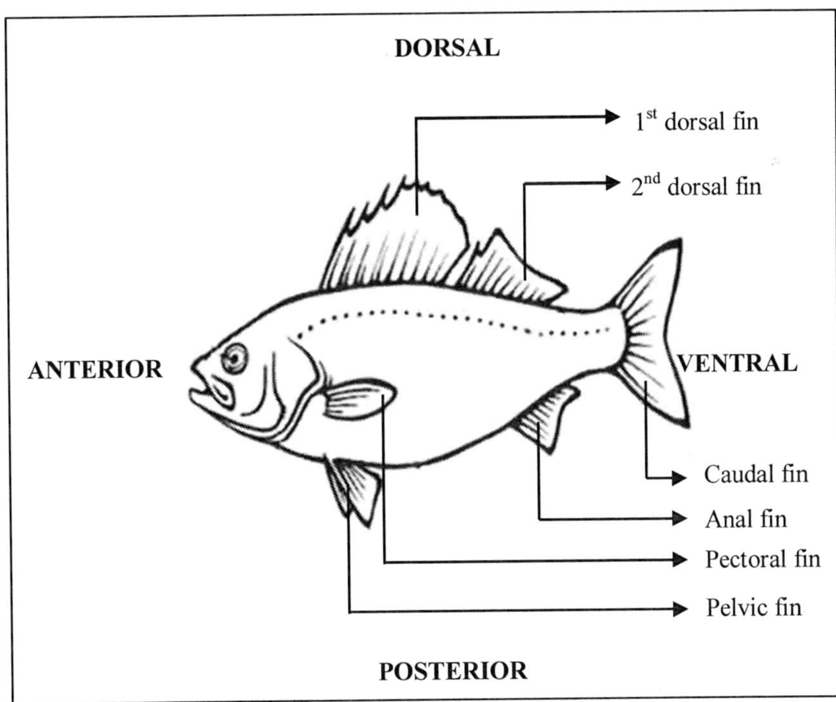

Fig. 2.1: Different paired and unpaired fins in fish.

4) **Caudal fin or Tail fin:** It is located at the end of the caudal peduncle and in most fish, the caudal or tail fin is the main propelling fin. Due

to the difference in the growth of the vertebrae in the two lobes of the tail, the caudal fin appears differently as described below:

a) **Protocercal:** It is possibly the most primitive type of caudal fin found in fish. In protocercal, the notochord also extends to the tip of the tail, but it is not expanded (Fig. 2.2). Example- Found in hagfish, lamprey, etc.

b) **Diphycercal:** It is externally similar to protocercal but found in fishes with a vertebral column. In this type, the vertebra extends to the tip of the tail and is expanded to a single large uniform lobe that lashes side to side. Example- Found in coelacanth, juvenile stage of modern bony fishes, etc.

c) **Leptocercal:** Structurally, it is similar to protocercal fin except here the vertebral column extends beyond the apex to form a long, symmetrical, cylindrical whip like structure. Example- Found in sting-rays.

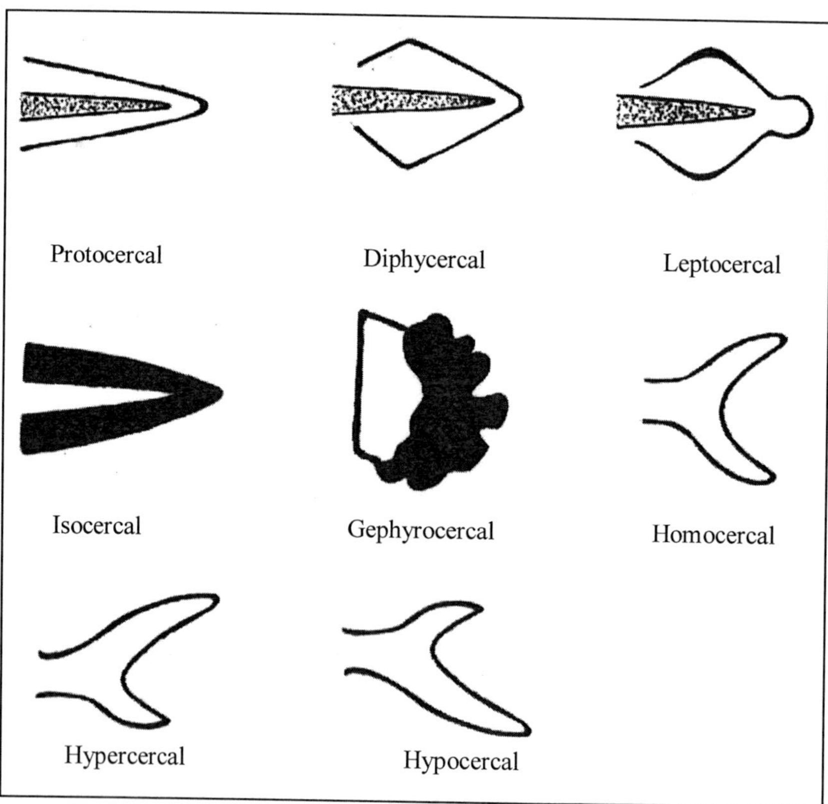

Fig. 2.2: Different types of caudal fins in fishes.

d) **Isocercal:** Homocercal tail lacking well-defined lobes is called isocercal tail and is formed by the fusion of caudal fin rays with posterior rays of dorsal and anal fins. This modification helps the bottom dwellers in swimming near the bottom of the sea. Example- Found in bottom-dwelling Holocephali, chimera, etc.

e) **Gephyrocercal:** In this type, the hinder portions of the dorsal and anal fins are fused and come over the aborted end of the vertebral column of a fish's tail. Found in fish species belonging to the family Molidæ.

f) **Heterocercal:** The vertebrae extend much either into the upper or lower lobe of the caudal fin making the caudal fin asymmetrical. It can be hypercercal or hypocercal.

If the vertebra is extended into the dorsal (upper) lobe making the dorsal lobe larger than the ventral (lower) lobe then it is called **hypercercal** tail. The larger dorsal lobe produces a lift force on the tail, and thus serves to direct the head of the fish down during swimming, a feature suitable for bottom-living fishes. Example- Found in cartilaginous fishes like sharks.

On the other hand, if the vertebra is extended into the ventral lobe making the ventral lobe larger than the dorsal lobe then it is called **hypocercal** caudal fin or tail. Due to this larger ventral lobe flying fish can produce a downward force on the tail the fish to swim upward in water to air for gliding. (Example- Found in flying fish). Thus, hypercercal tail simply helps the fish in propelling downward, while hypocercal tail helps the fish in propelling upward.

g) **Homocercal:** The vertebral column ends into a modified vertebra called the urophore complex, making a symmetrical caudal fin with equal sized upper (dorsal) and lower (ventral) lobe. Example- Found in Osteichthyes like rohu, catla, mrigal, etc.

2.1.3 Evolution of fins

The earliest known fish don't have paired fins, though some early jawless fish known to possess simple fin folds rather true fins along the sides of the body. Several theories have been put forward to explain the origin of fish fins. The three prevailing hypothesis that explaining the evolution of paired fins in fish are:

1) **Gegenbaur's gill arch theory (Gegenbaur 1870):** According to this hypothesis, paired fins are derived from gill structures. The posterior gill arches modified into pectoral and pelvic girdles, while the gill rays modified to form the skeletons of the fins (Goodrich 1906).
2) **Romer's fin spine theory (Romer 1966):** Spiny sharks (Class Acanthodii) possessed seven pairs of spiny appendages with attached fleshy web like membrane along their trunks. According to this hypothesis, two pairs of these spines (anterior and posterior) may have evolved into the paired fins, while the other spines may have been lost during the course of evolution.
3) **Lateral fin fold theory (Thacher and Balfour 1876):** It is now accepted as the possible explanation of the origin of paired fins. According to this theory, paired fins are budded from longitudinal, ventrolateral folds of the epidermis. These are extending backwards along the body from just behind the gills to the anus. The gradual suppression and degeneration of the intermediate portions of the folds lead to the formation of pectoral and the pelvic fins (Brand 2008). Figure

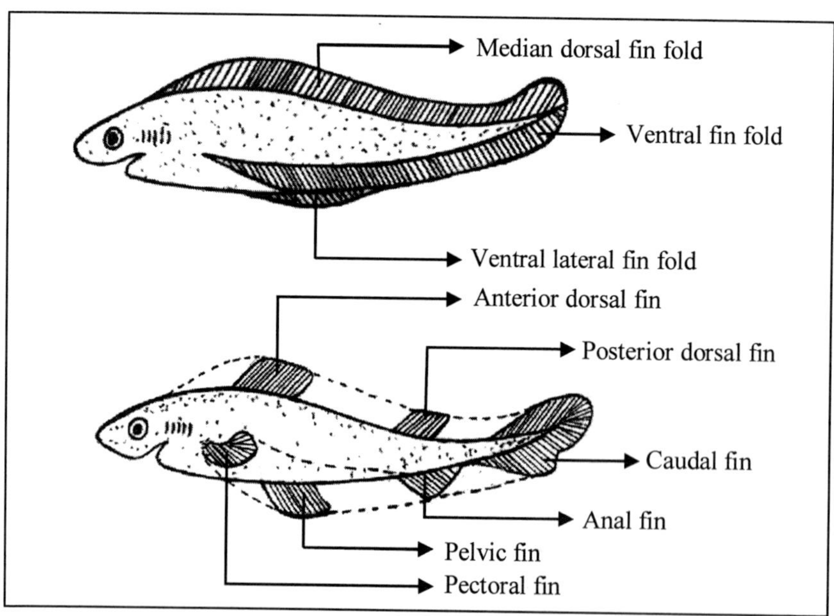

Fig. 2.3: Origin of fin in fishes according to Thacher and Balfour's lateral fin fold theory.

2.1.4 Modifications of fins in fishes

As a locomotory organs fins in most of the fishes help in propelling, steering and turning the body in water during swimming. In addition to this, many different species of fish have modified their fins to perform specialised functions in a particular habitat. Some of the remarkable examples where fishes show modification in their fins are described below

1) **Claspers and gonopodia for internal fertilisation:** The male cartilaginous fish (example- sharks and rays) modified the posterior portion of their pelvic fin into an intromittent organ called claspers. They used their claspers to deliver semen into the female's cloaca during mating. During mating, sharks raise one of the claspers to allow water entry into a siphon. They then insert their claspers into the cloaca of the female where the siphon begins to contract to expel water and sperm.

   Some male fish species of the family Anablepidae (Example- four-eyed fish) and Poeciliidae (Example- guppy) modified the third, fourth and fifth rays of the anal fin into a tube-like movable intromittent organ called gonopodium. During mating male shortly inserts its gonopodium into the female's cloaca and delivers the milt. As the sperm is preserved in the female's oviduct, females can fertilise themselves at any time without further help from males.

2) **Attracting males during courtship:** The females of species *Pelvicachromis taeniatus* possess modified large purple coloured pelvic fin. It has been found that male always prefers the female with a large pelvic fin as their mating partner (Baldauf et al. 2010).

3) **Modification of dorsal fin for temperature regulation:** Sailfish (*Istiophorus*) possess unique highly modified large sized erectile dorsal fin popularly known as **sail**. It has been found that this large dorsal fin is highly vascular. Thus, these blood vessels may be used for rapid heating and cooling of the body, according to the surrounding environment allowing the fish to swim faster and longer.

4) **Modification of dorsal fin for venom injection:** Fishes like reef stonefish, lionfish, scorpion fish, etc. are venomous fish. All of these possess modified spine bearing dorsal fins. The spines of the dorsal fins have venom glands at their bases, and thus it is used to inject venom into the body of prey.

5) **Modification of pectoral fins for walking:** Mudskippers possess highly modified large pectoral fins almost like small legs. Unlike

most fish, they use their pectoral fins to walk on muddy environments, land and under the water mainly to avoid predators.

Species of epaulette shark, *Hemiscyllium ocellatum* lives in shallow waters where swimming is difficult. They evolved large, muscular pectoral fins to walk over rocks and sand.

6) **Modification of pectoral fin for flying:** Fish species belonging to the family Exocoetidae modified their pectoral fins into a wing-like structure. This extremely long wing-like fins enables the fish in gliding flight for certain distances above the water's surface. The flying fish evolved this mechanism to escape from oceanic predators. These modified fins allow them to propel themselves out of the water at speeds of around 35 miles per hour and to glide for up to 200 metres.

7) **Modification of fins into sucker:** In many fishes, fins are found to be modified into a sucker for adhering to rocks and corals. For example, in fish goby, the pelvic fins are fused to form a single disc-like sucker.

   In remoras (also called suckerfish), the first dorsal fin is modified into an oval, sucker-like organ which allows them to attach themselves to the skin of larger marine animals like the whale, shark, etc. This allows them to feed on discarded food particles and faeces of the larger animals.

   In lumpfish, the pelvic fins are evolved into adhesive discs which allow the fish to adhere to the substratum.

8) **Modification of dorsal fin into fishing rod:** In anglerfishes (order Lophiiformes), the first spine of the dorsal fin is modified into a long filament called **illicium** tipped with a fleshy part called **esca**. The structure is similar to a fishing rod and lure. By wiggling the esca they attract their prey close enough to swallow them.

2.2 LOCOMOTION IN FISHES

The act of movement of an organism from one place to another by the action of appendages like fins to find food, a mate, a suitable niche, or to escape from predators is called locomotion.

Modes of locomotion found in fish: On the basis habitat, fishes perform different modes of locomotion such as swimming, flying, walking and burrowing:

A] Swimming: The most economical and prevailing type of locomotion in fishes is swimming in the surrounding water. This form of locomotion is economical in the sense that fish does not have to spend energy to counter gravity as nearly neutral buoyancy can be achieved either by the air bladder (bony fishes) or by the fat and oil accumulated in liver (sharks). For example, a salmon fish spends only 0.39 kcal of energy for swimming one kilometre distance, while on the other hand, ground squirrel spends 5.43 kcal for walking and flying gull spends 1.45 kcal of energy for flying.

Speed of swimming: The absolute speed at which a fish swims depend on both the body size and shape of the fish. In general, the larger fish swim faster than the smaller fish. For instance, a 30 cm sea trout can swim at a top speed of approximately 10.8 km per hour, while a 60 cm salmon can swim at a top speed of around 22.5 km per hour.

Richard Bainbridge (1958) formulated an equation for predicting the speed of swimming of fish using its size and frequency and amplitude of the tail beat. According to him,

V = 1/4 [L (3f-4)], where 'V' is the velocity in centimetres per second, 'L' is the length of the fish and 'f' is the frequency of tail beats per second. Thus, the tail beat is found to be directly related to the length of the fish and speed of swimming (Bainbridge 1958).

Kinds of swimming: Swimming can be classified on the basis of energy cost (Hoar and Randall, 1978), the body parts involved and the segment of body involved in generating thrust.

1) **On the basis of energy cost:**
   a) **Sustained swimming:** In this type swimming, speed is slow but the length of the swimming period is longer and is generally employed during long distance migrations. The principal source of energy is aerobic respiration.
   b) **Burst swimming:** In burst swimming, the speed is very high, but the fish swims for a shorter period and is generally employed for escaping from predators, chasing a prey or for swimming against the water currents. Energy is generated by anaerobic respiration.
   c) **Prolonged swimming:** This type of swimming has intermediate speed and energy utilisation. Both aerobic and anaerobic respirations supply the energy demands.

2) **On the basis of body parts involved:**
   a) **Axial-based locomotion (body and tail):** In this type of locomotion, fish propel them against the drag in the water primarily with the body and tail. Example- Found in sawfishes, guitarfishes and electric rays (Rosenberger 2001).
   b) **Pectoral-fin-based locomotion (pectoral fins):** In pectoral-fin-based locomotion, propulsion against the drag is executed with the pectoral fins. Pectoral fin based locomotion is of two types: undulation and oscillation (Rosenberger 2001).
      - **Undulation:** It is also known as rajiform locomotion and it is defined as having more than one wave present on the fins at a time. Example- Found in skates and most stingrays.
      - **Oscillation:** It is also termed as mobuliform locomotion and is defined as up and down movement of fins with less than half a wave on the fins. Thus, it actually involves spinning of the fins on the attachment point rather a wave like motion. Example- Pelagic stingrays, such as manta, cownose, and bat rays perform oscillation.
3) **On the basis of body segment involved:**
   a) **Sub-carangiform locomotion:** In this type of swimming, lateral undulatory waves are generally occurring nearly rear half of the fish's body. Example- Found in trout, salmon, etc.
   b) **Carangiform locomotion:** In carangiform locomotion, only the posterior one-third of the body bends with the passage of contraction waves. Example- Found in of the herrings (Family- Clupeidae), some characins (Family- Characidae), mackerels (Family- Scombridae).
   c) **Ostraciform locomotion:** In contrast to carangiform locomotion, in ostraciform locomotion, only the caudal fin oscillates from side to side. Example- Found in found in boxfish and trunkfish.
   d) **Thunniform locomotion:** In this type of locomotion, the fish moves due to sideways movement in the tail and the caudal peduncle. Example- Found in high-speed long-distance swimmers, like tuna, lamnid sharks etc.
   e) **Anguilliform locomotion:** Locomotion in which lateral undulation of the entire body occurs due to contraction of myotomes is called anguilliform locomotion. In this type of locomotion, the sinuous waves passing from head to tail cause

each segment of the body to swing laterally across the axis of locomotion, where each swinging segment describes a figure-eight loop. The disadvantage of this type of swimming is that it consumes a lot of energy as the whole of the body is involved in locomotory movements. Example- Found in long, slender fish like eels (Fig. 2.4).

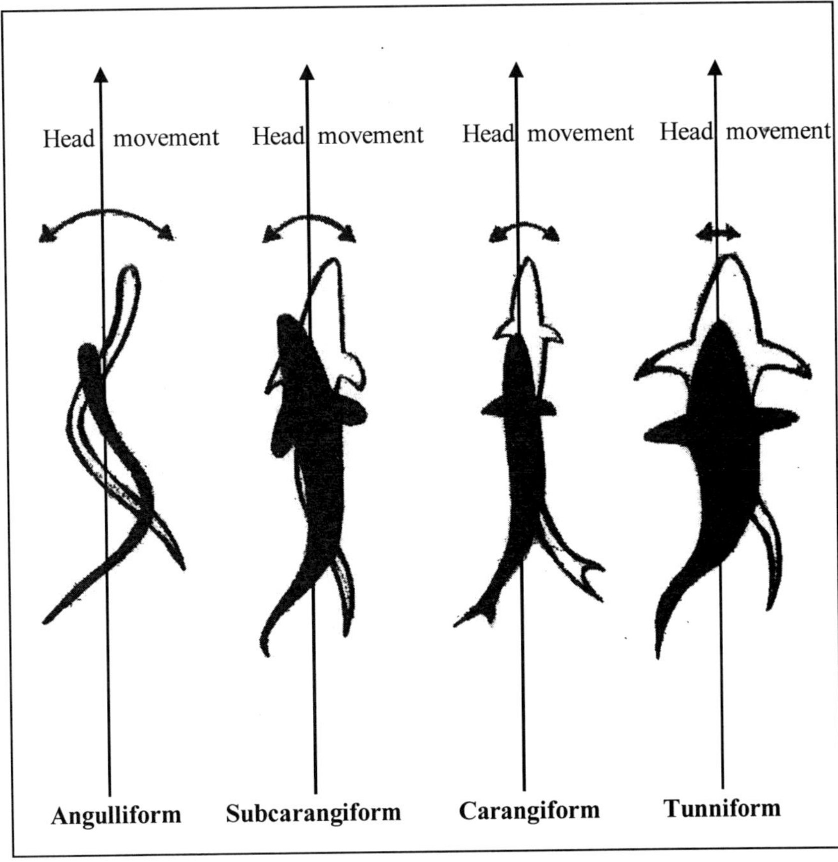

Fig. 2.4: Different kind of locomotion in fishes on the basis of body segment involved.

Exceptions to these kinds of swimming: Beside these five types of lateral body/caudal fin propulsion, there are some fishes which perform different undulatory movements during swimming. Some of these are discussed below:

- South American knifefish (Gymnotiformes) swim using undulations of their elongated anal fin, while keeping their body still. It is believed that it is an adaptation to keep generated electric field undisturbed.
- Fishes like species of batoids use pectoral locomotion exclusively to propel themselves through the water (Rosenberger 2001).
- Seahorse and pipefish (lack all fins except dorsal fin) use their single dorsal fin to push the body forward in a vertical position.
- Fishes like *Amia* use its long dorsal fin, while *Notopterus* and *Wallago* use their long anal fin to propel the body forward.

Role of fins in swimming: In general a) pectoral fins are used for balancing and braking b) pelvic fins assist the fish in going up or down through the water, turning sharply (steering), and stopping quickly c) dorsal fin prevent fish from side to side rolling and assist in sudden turns d) anal fin helps to maintain stable equilibrium and e) caudal fin is used for propulsion. Flying fish uses long ventral lobe of the heterocercal tail to swim upward in water to air for gliding, while bottom dwellers use long dorsal lobe of the heterocercal tail to propel them downward during swimming.

B] Flying: In addition to swimming, marine fish species belonging to the family Exocoetidae has evolved for gliding flight. They are commonly found in the epipelagic or sunlight zone in tropical and warm subtropical ocean. Their extremely long wing like modified pectoral fins enables them in gliding flight for certain distances above the water's surface. Species belong to the genus *Exocoetus* have only enlarged pectoral fins, while the species belong to the genus *Cypselurus* have enlarged pectoral and pelvic fins.

The flying fish evolved this mechanism to escape from oceanic predators like dolphins, tuna, marlin, etc. These modified fins allow them to propel themselves out of the water at a speed of around 70 km/h and to glide for up to 400 metres (Kutschera 2005).

C] Walking: Besides swimming and flying some fishes popularly known as walking fish, or ambulatory fish evolved them to travel over land for extensive periods of time. These fishes employ springing, snake-like lateral undulation, and tripod-like walking as a means of their locomotion. Some of the remarkable examples of walking fishes are as follows:

- Mudskippers possess highly modified large pectoral fins almost like small legs and they use their pectoral fins to walk on muddy environments, land and under the water. They can even climb mangroves up to certain heights.

- Non-amphibious walking fish like flying gurnard or helmet gurnard, batfishes (Family- Ogcocephalidae) and tripodfish can walk along the sandy sea floor with their pelvic fins.
- A rather and more remarkable mode of terrestrial locomotion is shown by the Indian climbing perch, which can live without water for 6 days. It uses sharp spines on the lower parts of its gill plates to hold onto the ground and pushes itself forward using pectoral fins and tail.

D] Burrowing: A number of fish species are well known for their burrowing behaviour in sand or mud. For example, true eels, moray eels, and spiny eels burrow into sand, mud, or amongst rocks. Tilefishes belonging to the family Malacanthidae generally prefer gravelly or sandy substrate and look for shelter in self-made burrows, caves at the bases of reefs, or piles of rock.

2.3 FISH HYDRODYNAMICS

Swimming animals must overcome the drag (force acting opposite to the relative motion of an animal in the surrounding fluid) by producing thrust to move forward in the aquatic environment. Fishes face two types of drag live in viscous medium and are:

1) **Viscous drag:** Viscous drag is created due to friction between the fish body surface and water. The magnitude of viscous drag is inversely proportional to surface area. As larval fishes have a larger surface area, they often experience higher viscous drag. Slime coat provides a smooth surface that allows laminar flow and reduces viscous drag in fishes.
2) **Inertial or pressure drag:** It is the pressure of water from all sides on the fish body due to the displacement of water. Bulky fishes replace more water as compared to smaller fishes, and thus they experience high pressure drag. Both streamlined bodies of fishes and their motion in a straight line reduce inertial drag.

Forms of swimming behaviour: On the basis of which body structures involved in thrust production, swimming behaviour can be categorised into following two types:

1) **Median-Paired Fin (MPF):** In this mode, forward thrust against the drag is executed either by the two pectoral fins or both anal and dorsal fins. Example- Observed in rays, skates, etc.
2) **Body-Caudal Fin (BCF):** In this mode, forward thrust against the drag is executed by generating undulatory waves that propagate down the body to the caudal fin. Example- Found in seen in eels, lampreys, etc.

Correlation of body shape and hydrodynamics: The body shape of a fish affects its thrust performance like the speed of swimming.

1) **Shape of nose:** When a pointed nose incline to the water flow, the water comes off from the body surface easily, while water is hard to come off when a round nose inclines to the water flow (Fig. 2.5). Thus, a fish with pointed nose experience higher drag force compare to fish with the rounded nose (Hirata and Kawai 2001).
2) **Streamlined body shape:** Pressure drag is minimised by the streamlined shape of the fish. The streamlined form of the fish body reduces pressure drag by reducing the magnitude of the pressure difference over the body, allowing smooth water flow from the nose to the tail (Fish and Hui 1991).

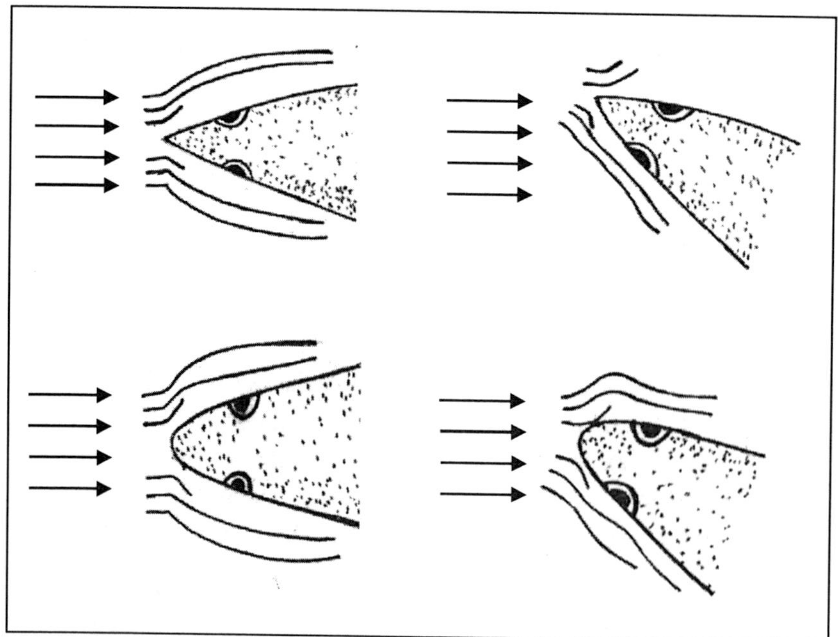

Fig. 2.5: Effect of nose shape on fish locomotion.

Relationships between swimming speed and caudal fin aspect ratio: Caudal fin plays a significant role in locomotory activities in fishes. The aspect ratio can be defined as the ratio of the square of caudal fin height to fin area. Thus, aspect ratio (A) of the caudal fin is equal to $\mathbf{h^2/s}$, where "h" is the height of the caudal fin and "s" is its surface area. High aspect ratio (shark) is a contributor to rapid swimming. On the other hand, fishes like trouts, minnows and perches can change aspect ratio of their flexible caudal fins according to their needs (Sherwood and Rose 2003).

Dynamic lift: To maintain depth most of the fishes either regulate the buoyancy by their gas bladder or store oils or lipids in their liver. On the other hand, fishes like sharks lack these features and they use the dynamic lift for this purpose. Similar to the use of wings of aeroplanes and birds, shark positioned their pectoral fins to create lift allowing it to maintain a particular depth in water.

Hydrodynamic versatility: Fish fins exhibit considerable hydrodynamic versatility with respect to the situation. A lateral push of the caudal fin on the water produces two reactive forces, namely the forward thrust and lateral force on the opposite side at right angles to the axis of the body. Forward thrust propels the body forward, while the lateral force makes the head yaw (twisting about a vertical axis.) from side to side.

1) **Yawing:** It is the side to side movement of the head due to lateral reaction force. Yawing is countered by pectoral fin, allowing a fish to move in a straight line.
2) **Pitching:** The up and down movement of the head produced either due to uneven drag on the body or by heterocercal or hypocercal tail fin is called pitching. Pitching is countered by pectoral fins.
3) **Rolling:** Twisting of the body on its anterior-posterior axis is called rolling. Rolling is controlled by the dorsal fin when a fish wants to turn right or left.
4) **Steady swimming:** In steady swimming fishes like bluegill sunfish recruit multiple fins (dorsal, pectoral, and caudal fins) to generate thrust.
5) **Slow turning:** Here, the total lateral force is executed between the strong-side pectoral fin (near to the turning stimulus) and the dorsal fin.

During turning the strong-side pectoral fin executes lateral force to rotate the head away from the stimulus, while the weak-side pectoral fin generates posteriorly directed thrust to cause body translation (Lauder and Drucker 2002).

2.4 FISH SCALE AND ITS TYPES

2.4.1 Scales

Fish scales are microscopic to large rigid armour external protective structure found on the skin of most fishes (exception- electric ray, catfish, etc.). Fish scales are part of their integumentary system and are mesodermal derivatives. Among different species of fishes, the scales show high diversity in shape, size and structure.

For example, saltwater finfish tuna, common eel possesses tiny, microscopic scales, *Chela* (family- Cyprinidae) possesses small sized scales, while the scale length of the Indian Mahseer (*Tor tor*) is known to reach over 10 cm. Diversity in scale type is also found in a single fish as well as within a species between the male and female (Example- flatfishes).

2.4.2 Types of scales found in fishes

The scales show high diversity in their shape and structure. The different types mesodermal derived scales identified in fishes are as follows:

1) **Placoid scales:** These are tooth-like scales and have one median spine and two comparatively small lateral spines. Placoid scales are structurally homologous (possess central pulp cavity, layers of dentine and enamel-like substance called **vitrodentine**) with vertebrate teeth, and thus they are also called dermal denticles.
   Example- Found in cartilaginous fish like sharks, Torpedo, Dogfish, etc.
2) **Leptoid scales:** Leptoid scales are common among higher-order bony fish like teleosts. These scales are found in overlapping positions on each other on skin. It is of two types:
   a) **Cycloid scales:** They have a smooth and uniform outer margin (Fig. 2.6). Example- Found in Salmon, Rohu, Catla, Mrigel, etc.

b) **Ctenoid scales:** They have a toothed outer edge and unlike ctenoid scales, they possess tiny teeth called **ctenii** along their outer edges. The principal chemical constituents include hydroxyapatite, calcium carbonate and collagen. Example- Found in Perch, Bass, Crappie, etc.

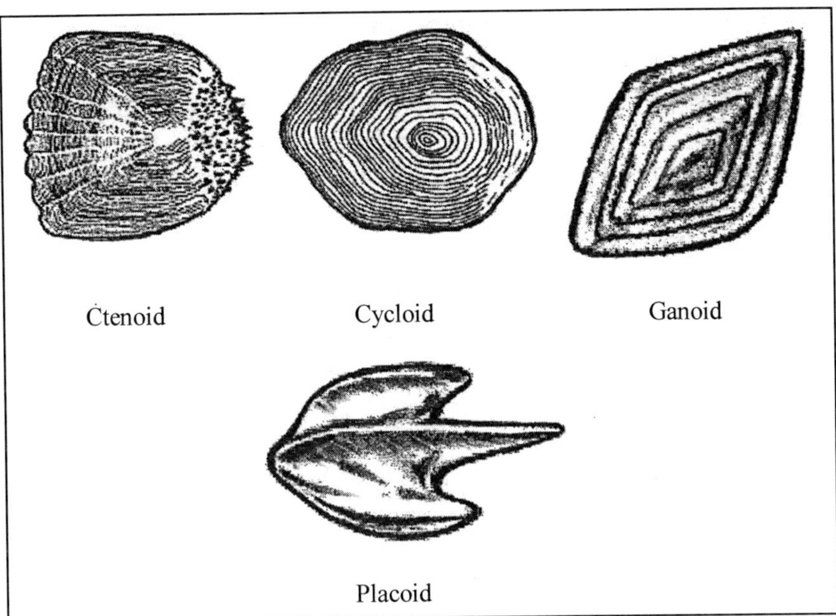

Fig. 2.6: Different types fish scales.

3) **Cosmoid scales:** Cosmoid scales have a hard, enamel-like outer layer called vitrodentine, an inner layer of dentine-like material called cosmine and then the innermost layer of vascular bone called isopedine.
   Example- Found in ancient lobe-finned fishes, lungfishes, etc.
4) **Ganoid or rhomboid scales:** Theses are diamond-shaped, shiny and non-overlapping hard scales. These are similar to cosmoid scales except a layer of hard inorganic salt substance called **ganoin** lies above the cosmine-like layer and under the enamel (Lagler et al. 1962).
   Example- Found in Paddlefishes, Gars, Bowfin, etc.
5) **Elasmoid scales:** These types of scales are composed of a dense layer of isopedine and a reduced layer of dentine.
   Example- Found in Coelacanths, Amiids, etc.

2.4.3 Modifications of scales in fishes

The primary purpose of scales is to give the external protection. However, to serve various functions according to their habitat, different groups of fish have evolved a number of modified scales. Some of the modified scales are discussed below:

1) **Modification of scale into spine:** In many fish, scales are modified into spines for own protection.
   For example, porcupinefish commonly called blowfish modified their scales into long spines. These spines normally lie flat against the body, but when porcupinefish frightened, it quickly swallows lots of water, which makes the fish expand like a balloon. As the porcupinefish puffs up, all the spines stick straight out, making it even harder to eat by its predator. Similarly, in Surgeonfishes (Acanthuridae) scales are evolved into weapon-like spines on either side of the tail called **thorn tails.**
2) **Deciduous scales for escaping from predators:** Deciduous scales are those scales which can be easily shed or rubbed off from the skin. For example, herring (Family Clupeidae), halfbeaks (Family Hemiramphidae), etc. possess deciduous scales. In these fishes, deciduous scale helps them in escaping from predators.
3) **Modification scales along lateral line:** In most bony fishes the scales found along the lateral line possess a central pore. These pores allow water to enter through scales to contact the mechanoreceptors (modified epithelial cells called hair cells) to detect movement and vibration in the surrounding water.
4) **Modification of placoid scales into saw teeth:** The stinging spines found in the tail of stingrays and the "saw" teeth of sawfishes and sawsharks are modified dermal denticles (placoid scales).

2.5 DETERMINATION OF FISH AGE

The exact age determination of fish is an important step in fishery science, particularly for stock assessments, developing proper management or conservation plans, artificial breeding, etc. As there are variations in body size among the fishes of particular species, it is difficult to correlate fish age with fish size.

Determining a fish's age is usually done with structures, which show age dependency growth. Following are the different structure of fish commonly used in determining the age of fish.

1) **Scales:** Fish scales show a unique growth pattern. During their growth, they produce small circular growth rings called **circuli** around them. The places where circuli are placed close together, a dark ring is formed called **annuli.** The scales of some species show bands of uneven seasonal growth, grows faster in summer, while slower in winter. The age of a particular fish can be determined by counting the number of annuli (rings) present on its scale.

   The technique simply involves the collection of scale, its washing and observation under the microscope. However, during collection, it should keep in mind that scales are collected from the same area on the same side of the fishes that age to be determined. It is suggested that scales found on the shoulder of the fish between the head and the dorsal fin are best for determining the fish age. The technique can be summarised into the following three steps:

   a) After washing off the fish under cold running water, a scale can be taken without making any damage in a head-to-tail direction using a forceps.
   b) The collected scale should be cleaned by dipping and rubbing between thumb and forefinger in distilled water to remove any mucus present.
   c) The scale can be then simply mounted on a glass slide and dry in the moderate heat of approximately 37^0C followed by counting of annuli under a microscope.

2) **Otoliths:** Otoliths are commonly referred to as fish ear bones or earstones. They are hard, calcium carbonate rich structures located behind the brain of teleosts. Its main function is the balance, orientation, and sound detection similar to the inner ear of mammals.

   Three pairs of otoliths are found in each fish which includes one large pair called **sagittae** and two small pairs called **lapilli** and the **asteriscii.** Due to its large size compare to the other two, sagittal otolith is used for age determination of most fish by counting the annual growth rings on the otolith. The shape of otoliths is species specific. In general, different fish species have differently shaped otoliths. If the otolith is thin enough (<300 mm), then a whole mount of the otolith can be used for counting of annuli. If the otolith is thick, it must be sectioned to analyse it more clearly. Prior to its observation, light burning of the otolith section is often preferred to make the annuli visible.

3) **Other bony structures:** In addition to scales and otolith, other bony structures used for age determination include vertebrae, opercula, fin rays, pectoral spines, etc. Some prefer examination of fin rays and pectoral spines as this method can be used without sacrificing the specimen.

2.6 GILL AND ITS ROLE IN GASEOUS EXCHANGE

2.6.1 Gills

Gill is the respiratory organ of many aquatic animals, including most of the fishes (except lungfish or salamanderfish belonging to the subclass Dipnoi) that exchanges oxygen and carbon dioxide between water and blood. Gills are found on either side of the pharynx and are composed of vertical rows of filaments which possess comb-like processes called gill lamellae. The lamella increases the surface area for gaseous exchange (Fig. 2.7).

1) **Cartilaginous fish:** The respiratory system of cartilaginous fish like *Scoliodon* is composed of five pairs of gill-pouches containing lamelliform gills, on either side of the lateral wall of the pharynx. Each gill-pouch opens into the cavity of pharynx through a large internal branchial aperture and to the exterior through a narrow external branchial aperture commonly called gill slit. The adjacent gill pouches are separated from each other by a fibro-muscular partition called gill septum, which remains anteriorly supported by cartilaginous gill arc. The endodermal mucous membrane of each gill-pouch is raised into a series of horizontal folds called gill lamellae. The branchial lamellae are richly supplied with blood capillaries for gaseous exchange.

   On the basis of presence of lamellae, the gill arc can be of two types- a) a gill arc is called hemibranch or demibranch when gill lamellae are present on only one face of the septum and b) a gill arc is called holobranch when gill lamellae are present on both anterior and posterior faces. In *Scoliodon*, the hyoid arc possesses a posterior hemibranch; the first four branchial arcs possess holobranch (Fig. 2.8), while the fifth branchial arc has no branchial lamellae. Thus, *Scoliodon* has nine hemibranches on each side. The posterior hemibranch of each holobranch is comparatively larger than the anterior one.

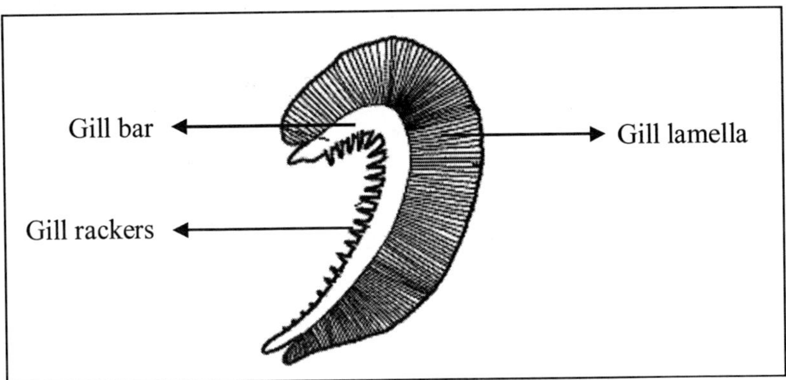

Fig. 2.7: Structure of gill lamella.

Each interbranchial septum extends much beyond their respective gill lamellae to form a flap-like protective structure for gills and gill slits. In most Elasmobranchs, the spiracle (openings on the surface of some animals, which usually lead to respiratory systems) present between the hyoid arches on each side bears minute branchial lamellae and opens to the exterior by an external branchial aperture just behind the eye. It is a pseudobranch or false gill as it is supplied with arterial blood and plays no part in respiration. But in *Scoliodon*, the spiracles are vestigial pits in the pharynx as it has no gill lamellae and external opening.

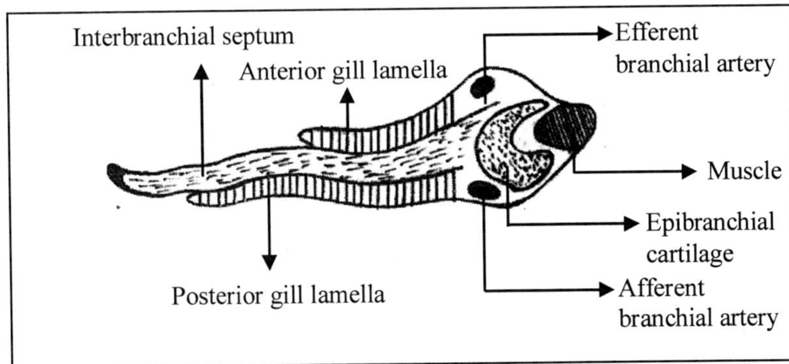

Fig. 2.8: A holobranch of dogfish.

2) **Bony fish:** In bony fish, the gills lie in a branchial chamber, covered and protected by a hard bony flap called operculum. The operculum plays a significant role in bony fish by adjusting the water pressure inside the pharynx. The opening of operculum causes pressure to drop which results in water flow towards the lower pressure over the gill lamellae for gaseous exchange. Most of the bony fish species have five pairs of gills below the operculum. Unlike cartilaginous fish, the gill arches of bony fish have no septum.

2.6.2 Role of gills in gaseous exchange

Gill lamellae are the site of gaseous exchange between water and blood in fishes. Two types of lamellae are found in fishes. The lamellae come out of the interbranchial septum are called primary lamellae and that come out of the primary lamellae are called secondary lamellae, the purpose of both is to provide a large surface area: volume ratio.

The alternate opening of the mouth and opercula helps to force water across the gill lamellae. The fish opens its mouth to let water in, then closes mouth and forces the water over the gills at regular intervals to the outside aquatic environment through either gill slits (Chondrichthyes) or operculum (Osteichthyes). However, some larger fish like tuna swim continuously with their mouth open to keep water flowing across their gill surface. Thus, water flows through the gills in one direction and it is very important for fish because of the low oxygen concentration in water compared to air. During this process, when water comes in contact with gill lamellae, dissolved oxygen from incoming water diffuses into the blood in the capillary present in gill lamellae, while carbon dioxide diffuses from blood to water which is then passed out from the body through outgoing water current (Fig. 2.9).

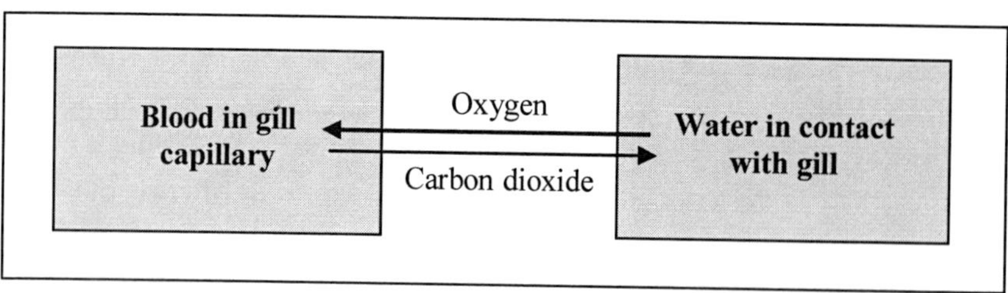

Fig. 2.9: Direction of gaseous exchange in brachial respiration.

Advantages and limitation: The advantages and limitation of gaseous exchange between water and blood at gill are summarised below:

1) The presence of primary and secondary lamellae provides a large surface area for gaseous exchange. Moreover, flowing of water across the lamellae keeps them apart, and thus maximising the surface area further.
2) The counter-current system is an advantage of the gas exchange system in bony fish. In counter-current flow, the deoxygenated blood in blood capillaries of lamellae flows in the opposite direction to the flow of oxygenated water. It allows maximum oxygen to be absorbed by the blood from the water, by maintaining the concentration gradient (Fig. 2.10).
3) In addition to this, the unidirectional flow of water is another advantage as the gas exchange medium is replaced fully. This unidirectional flow of water allows oxygen to be absorbed from all of the water that comes across the fish gills, and thus increases the efficiency of gaseous exchange.

The only limitation of this gas exchange system is that the fish can only live in water. In air, the filaments and lamellae will stick together drastically, and thus reducing the surface area to volume ratio, which in turn decreases the efficiency of diffusion of respiratory gases.

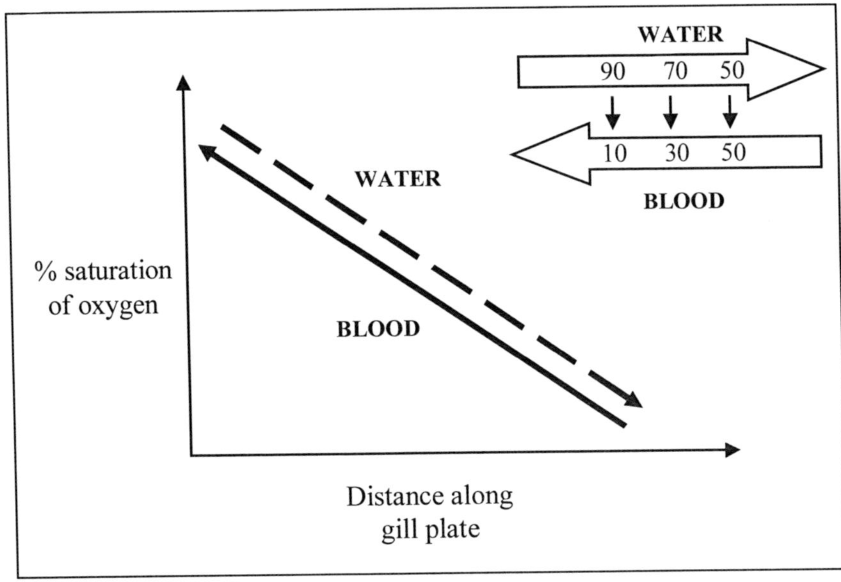

Fig. 2.10: Effect of counter current flow in gaseous exchange in gill.

2.7 SWIM BLADDER, ITS TYPES AND ROLE IN RESPIRATION AND BUOYANCY

2.7.1 Swim bladder

Swim bladder, also called air bladder is a white glistering, non-rigid, thin walled, gas-filled sac located against the roof of the body cavity in most of the bony fishes. It is derived from an outpocketing of the digestive tube. It mainly functions as a hydrostatic organ, which enables the fish to maintain its water depth without floating upward or sinking and without expending much amounts of energy. The principal gases found in swim bladder include O_2, CO_2 and N_2.

Structure of swim bladder (teleost): The higher bony fishes possess the most specialised form of air bladder, where it primarily functions as a hydrostatic organ. It consists of two gas-filled sacs located in the dorsal portion of the fish. The walls of the bladder are lined with guanine crystals to make them impermeable to gases.

The gas bladder of can be divided into a comparatively large a) gas producing anterior chamber and a small b) gas absorbing posterior chamber. The wall of the anterior chamber contains gas gland or red gland, which converts glucose to lactic acid and also produces carbon dioxide, which is then diffused into blood circulating around the bladder. The resulting acidity in venous capillaries causes dissociation of oxygen from haemoglobin (called **Root effect**) of the blood, which then partly diffuses into the swim bladder. The blood then first enters a **rete mirabile** (a network of arteries and veins lying very close to each other with the countercurrent blood flow) where all the oxygen diffuses back from venous capillaries to the arterial capillaries supplying the gas gland. This increases the partial pressure of oxygen in the arterial capillaries allowing oxygen to diffuse into the swim bladder via the gas gland. The oval gland of the posterior chamber allows reabsorption of gases by blood.

Evolution of swim bladder: The close similarity in origin, structure and development between swim bladders and lungs clearly suggested both are homologous. It is believed that the primitive, simple sacs found in early bony fishes as an adaptation to gulp air under oxygen-poor conditions in the Devonian period undergo three lines of evolution (Fig. 2.11):

1) **Evolved into lungs:** They evolved into lungs of today's land vertebrates and some fish like lungfish, garfish, and bichir.

2) **Evolved into swim bladder:** In another line, they evolved into either physostomous (in carps, trouts, catfish, and eels) or physoclistous (in perch, cod) type of swim bladder.
3) **Lost in elasmobranchs:** These primitive, simple sacs were completely lost in some organisms like Elasmobranches (e.g., sharks and rays) and many marine bony fishes. They lack both lungs and swim bladders. However, as an alternative, these fishes evolved with heterocercal and pectoral fins, which provide the necessary lift.

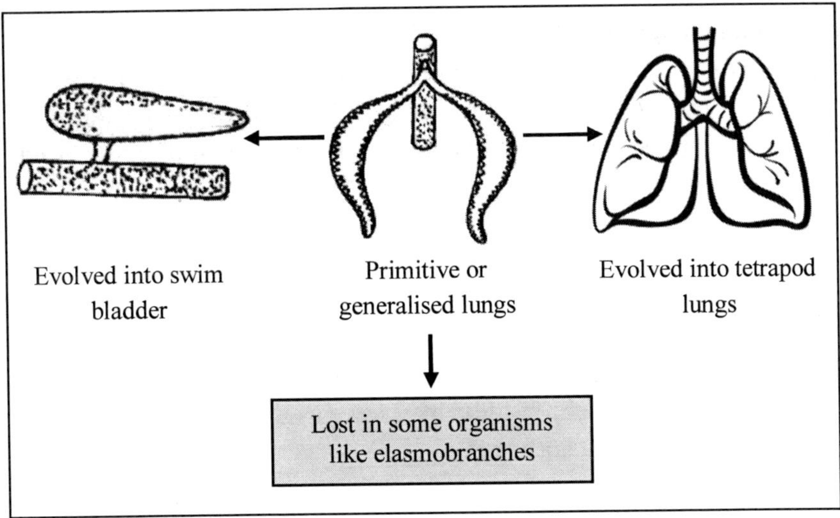

Fig. 2.11: Three lines of evolution from primitive or generalised lungs in vertebrates.

Advantages of swim bladder (function):

1) **Hydrostasis:** Swim bladder enables the bony fish to maintain a stationary level in water without expending much amounts of energy in swimming.
2) **Respiration:** In most primitive fishes, the air bladder serves as a supplementary respiratory organ or lungs.
3) **Oxygen reservoir:** During emergency condition, physoclists and physostomes often use oxygen present within their bladder as an emergency backup.
4) **Audition:** In many freshwater species like common carp and wels catfish the swim bladder remains connected to the labyrinth of the inner ear by the Weberian apparatus. The presence of inner ear- swim bladder connections increases the hearing ability of this fishes.

5) **Sound production:** In many fishes (goliath grouper, black drum, toadfish, carapid fishes, silver perch, etc.), contraction of sonic muscles attached to their swim cause the swim bladder to contract and expands at a rapid rate; thus, creating high-frequency drumming sounds (45-1000Hz). They produced drumming sound to startle the predators or competitors, to attract mates, or as a fight response. In addition to this, sound may also be produced by the vibration of the incomplete septa present inside the swim bladder.

Disadvantages of swim bladder:

1) One major disadvantage of the swim bladder is that neutral buoyancy can be achieved only at a narrow range of depth. Beyond this range
   a) If a fish swims below its buoyancy range, it will have to expend much energy in order to keep them from sinking.
   b) If a fish swims above its upper buoyancy level, it becomes exceedingly buoyant, making the fish out of control.
2) A much greater depth flexibility and speed in moving through water columns is found among fishes which do not have a swim bladder.
3) The presence of swim bladder might enable predators to more easily locate the fish as swim bladder may serve as an acoustical target and the sounds can be bounced off.

2.7.2 Types of swim bladder

In general, there are two types of swim bladder 1) physostomous (open type) swim bladder and 2) physoclistous (closed type) swim bladder (Fig. 2.12).

1) **Physostomous:** In this type, the swim bladders remain connected to the gut by a pneumatic duct, which allows the fish to fill up the swim bladder by gulping air through the mouth. This gulped air passes into the gut and then they will force it into the swim bladder to fill it up. The air is stored in the swim bladder to counteract the gravitational force and to give the fish neutral or nearly neutral buoyancy. Example- Found in carps, trouts, catfish, eels, lungfish, etc. On the basis of number of the lobes and site of the pneumatic duct connection, following forms of swim bladder are found in physostomous fishes:

- **Unequal bilobed sac:** In *Polypterus* and *Calamoichthys*, the air sac is a bilobed one (unequal lobes) and it remains connected ventrally to the oesophagus by a single duct.

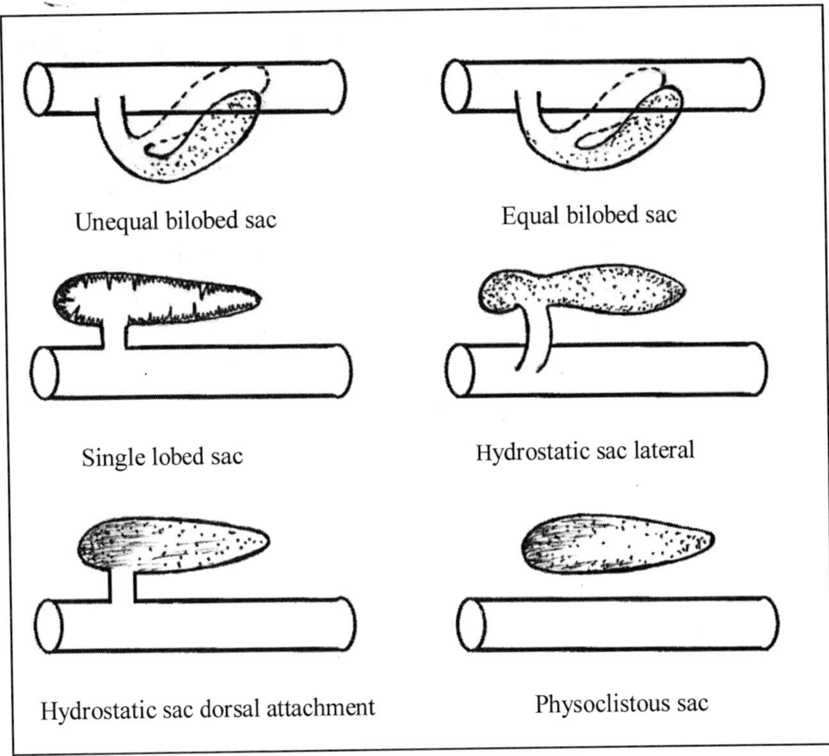

Fig. 2.12: Different types of swim bladder found in fishes.

- **Equal bilobed sac:** Lungfishes also possess a bilobed air sac with alveoli and septa but the two lobes are equal in size.
- **Single lobed sac:** Fish species of *Amia* and *Lepidosteus* possess a single large highly vascular lobe, which remains connected dorsally to the oesophagus, and thus serves both as respiratory and hydrostatic organ.
- **Single lobed hydrostatic sac with lateral attachment:** In teleost like *Erythrinus*, the air bladder becomes a hydrostatic organ, which remains connected laterally to the oesophagus.
- **Single lobed hydrostatic sac with dorsal attachment:** In the more generalised group of teleosts, the air bladder is almost similar to the previous one except in this pneumatic duct connect the air sac dorsally to the oesophagus.

2) **Physoclistous:** In physoclistous type, there is no connection between the gas bladder and the alimentary canal. Thus, it gulps air from the atmosphere to adjust their buoyancy. Instead, the gas gland present in these fishes allows diffusion of oxygen, carbon dioxide, and nitrogen between blood and swim bladder. Example- Found in specialised teleosts like perch, cod, etc.

2.7.3 Role of swim bladder in respiration and buoyancy

The swim bladder mainly serves as an accessory organ of respiration or lung or as a hydrostatic organ to control buoyancy. In addition to these two functions, fishes also used the air bladder to produce drumming sound and to increase their hearing ability.

1) **Respiratory function of swim bladder:** In most primitive bony fishes like ganoids, lungfishes, holosteans and in spiny-rayed teleosts, the air bladder serves as the lungs and as accessory respiratory organ respectively. The fish species of *Polypterus* has a pair bilobed sac of (primitive lungs) allowing the fish to periodically gulp air from the surface of the water. During the course of evolution, respiratory organs like the lung of lungfish and vascularised air bladder of holosteans are evolved from the primitive air bladder.

   Holostean like *Amia* possess both a well vascularised air bladder and efficient gills. It is an almost exclusively water breather at 10^0C but with increasing surrounding temperature (30°C) and activity, it starts taking fresh air using its swim bladder. At 30°C, three times more oxygen is taken from the air than from water by swim bladder, but the gills continue to be the principal site for CO_2 elimination (Johansen et al. 1970).

   In *Notopterus*, the gills are highly reduced, while the air bladder is highly developed allowing the fish to live without water for few hours. In the trunk region, air bladder develops many (14-15) blind pouches with finger-like projections. These projections are vascular and serve as the site for gaseous exchange. Thus, in this fish, air bladder serves as a true lung, making the fishes nearly a complete air breather (Munchi and Hughes 1992; Dehadrai 1962). During the active intake phase, fresh air is forced from buccopharynx to air bladder.

In addition to this, during emergency conditions, both physoclists and physostomes often use oxygen present within their bladder as an emergency backup.

2) **Hydrostatc function of swim bladder:** In higher bony fishes or teleosts, the air bladder is dorsal in position and the internal structure is not alveolar, suggesting that in these fishes it functions chiefly as a hydrostatic organ. The dorsal position of the swim bladder keeps the centre of mass below the centre of volume, making it an excellent hydrostatic organ.

The swim bladder works by varying the amount of gas within it, giving the fish a neutral or nearly neutral buoyancy. The rising and sinking of most bony fishes in the water can be compared to the rising and sinking of a helium-filled balloon in the air. We know density is inversely proportional to volume.

- During ascending or rising in water a fish reduce its overall density by increasing its volume without increasing its mass by filling the swim bladder with gases. Due to an expansion of the swim bladder, fishes displace more water experiencing a greater force of buoyancy and the fishes are pushed to the surface.
- On the other hand, to descend or sink to the ocean floor, the fish completely deflated its swim bladder by releasing the gases. In this condition, fish has minimum volume (much denser) and it sinks to the ocean floor.
- To stay at a particular level, a fish must fill its swim bladder with gas up to the point at which it displaces a volume of water that is equal to the weight of fishes. As a result, the force of buoyancy and gravity cancel each other (neutral buoyancy) and the fish stays at that specific depth without any muscular effort.

Thus, swim bladder helps the bony fishes in its immersion and emersion as well as to stay at a particular water depth, without wasting much energy in continuous swimming. On the other hand, fishes which have no swim bladder like mackerels, sharks, and rays spend a lot of energy during constant swimming in order to keep from sinking. Fishes those lack swim bladder either use their pectoral fin and heterocercal tail to create lift or store fats or oils in the liver with a fixed density less than that of seawater.

2.8 OSMOREGULATION IN FISH

2.8.1 Elasmobranchs

Fishes belong to the subclass Elasmobranchii of class Chondrichthyes or cartilaginous fish are collectively called Elasmobranchs. Members of this subclass possess unique characteristics like-

a) Five to seven pairs of gill clefts opening individually to the exterior.
b) Rigid dorsal fins and small placoid scales on the skin.
c) The posterior portion of male's pelvic fins is modified into an intromittent organ called claspers for the transfer of sperm.
d) The swim bladder is absent and buoyancy is mainly maintained by large oil-rich livers.
e) Unlike the members of another subclass Holocephali, their upper jaw is not fused with the cranium.
   Example- sharks, rays and skates.

2.8.2 Osmoregulation and its importance

Osmoregulation is the physiological process by which an organism maintains salt and water homeostasis of the body.

Osmoregulation is usually achieved by excretory organs that serve also for the disposal of metabolic wastes. Thus, urination is a mechanism of both waste excretion and osmoregulation. Organelles and organs that carry out osmoregulation include contractile vacuoles of protists, nephridia, antennal glands, and malpighian tubules of invertebrates, and salt glands and kidneys of vertebrates.

Osmoregulation prevents body fluids of an organism in becoming too diluted or too concentrated. Thus, it helps in maintaining a constant osmolarity of the body fluids.

2.8.3 Mechanism of osmoregulation in elasmobranchs

Fishes live their lives completely surrounded by water on all sides. It surrounds them externally in their habitat and also comprises much of the body fluid. Fish must, therefore, use some sort of balance between these two separate and very different water environments. This balance is met through the processes of osmoregulation. Most of the elasmobranchs are marine, although around 10% are estuarine, 2% are euryhaline and 1% is obligate in fresh water (Martin 2005).

Problems: In the marine water, the body fluid of fish, including the elasmobranchs has a lower concentration of salt compared to the external environment. Thus, the body fluid of the marine water fish is hypotonic to its external watery environment, and thus faces the following problems-

1) As sea water passes through the mouth and over the gill membranes, water molecules diffuse out of the blood into the sea water by osmosis. Therefore, marine fish must have a mechanism for getting or conserving water.
2) Drinking sea water brings a large quantity of salt into the blood and marine fish must have a mechanism for removing this excess salts.

Solution or mechanism of osmoregulation: Unlike the other marine animals, elasmobranchs perform osmoregulation mainly by regulating the concentration of urea and other body fluid solutes (TMAO) and excretion of salts by the rectal gland.

1) **Osmoregulation by producing metabolic urea:** All elasmobranchs (exception freshwater potamytrygonid rays) are ureotelic as they excrete urea as the main excretory product. These organisms accumulate organic nitrogenous compounds, such as urea and trimethylamine oxide (TMAO) in their body fluid, making the body fluid nearly isotonic to sea water. Over 30% of total plasma osmolarity is due to urea (Hammerschlag 2006).

   Elasmobranchs (exception potamytrygonid rays) have an active ornithine cycle in the liver, which synthesises the serum urea. In addition to this, the kidney tubules of marine and euryhaline elasmobranchs in seawater are capable of reabsorbing most of the urea (70–99%) by the active transport of urea from the filtrate to body fluid.
2) **Salt regulation by rectal gland:** Marine and euryhaline elasmobranchs (sharks, rays, and skates) due to their hypotonic nature face the problem of the continuous diffusion of salts (Na^+ and Cl^-) into the body fluid from the external hypertonic sea water. In this organism, the rectal gland functions as a salt secretor for excreting excess salts from body fluid. The mitochondria-rich cells of the rectal glands possess Na^+/K^+-ATPase, which pump salt from the blood into the gland, from where it is excreted as a concentrated solution.

3) **Salt regulation at the gills:** In freshwater elasmobranchs (*Dasyatis, Pristis,* etc.), gill is involved in salt ion uptake. The Na^{+}/K^{+}-ATPase pump in the gill lamellae uptake ions from water into the body fluid. Besides sodium and chloride, salts which are absorbed by the cells present at gill lamellae are Li Na, Ca, Cl, Br, SO_4 and PO_4.
4) **Kidney salt excretion:** Elasmobranchs have the capacity to alter kidney function with respect to environmental salinity. When elasmobranchs are adapted to dilute environments (hypotonic), urine flow rate increases 20 to 50 times resulting in increased excretion of Na^{+}, Cl^{+}, Mg^{2+}, SO_4^{2-} and urea. On the other hand, in a saltwater environment, sodium, urea and chloride ions are reabsorbed and produce concentrated urine (Hammerschlag 2006).

2.8.4 Mechanism of osmoregulation in freshwater fish

Freshwater fish like rohu, catla, mrigal, etc. live their lives completely surrounded by water on all sides. It surrounds them externally in their habitat and also comprises much of the body fluid. But, fish do not always find themselves in isotonic environments. Body fluids of freshwater fish are hypertonic to their surrounding environment and therefore continuous uptake of water takes place in freshwater fish through the gill membranes into the blood. Fish must, therefore, use some sort of balance between these two separate and very different water environments that is body fluid and the external aquatic environment. This water/salt balance is met through the processes of osmoregulation.

Problems: In fresh water, the body fluid of fish has a higher concentration of salt compare to the external environment. Thus, the body fluid of freshwater fish is **hypertonic** to their external water environment, and thus they face the following problems-

1) As fresh water passes over the gill lamella, water moves from the fresh water into the blood by osmosis. Thus, these fish must produce a very large volume of urine to balance this large intake of water (counteract the gain).
2) During the excretion process, they lose plenty of essential ions (Fig. 2.13). Thus, these fish must have an efficient mechanism to overcome this ion loss (counteract the ion loss).

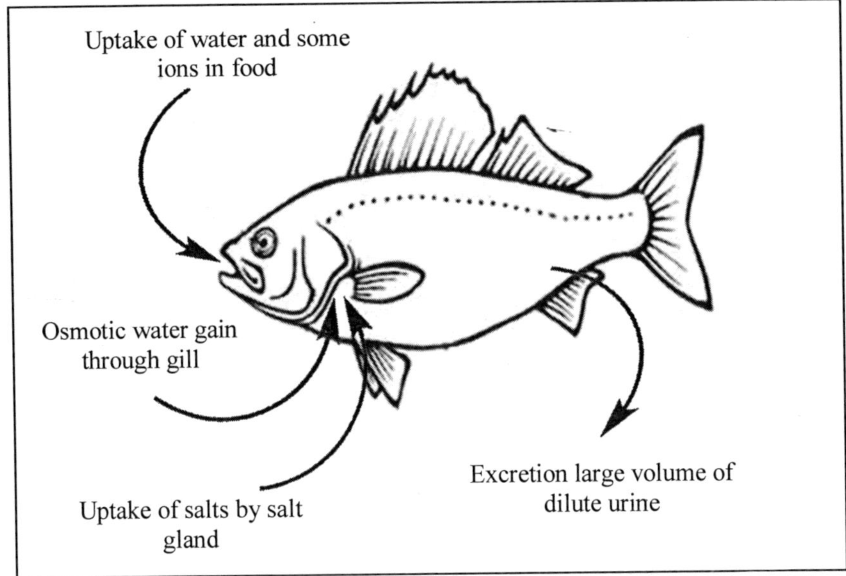

Fig. 2.13: Mechanism of osmoregulation in freshwater fishes.

Solution/ osmoregulation: In freshwater fish kidneys and gills perform the function of osmoregulation in the following ways-

1) **Excretion of large volume of diluted urine:** Freshwater fish has a highly developed kidney with a large number of glomeruli. The continual uptake of water in freshwater species is balanced by constantly producing large amounts of dilute urine.
2) **Active uptake of ions through gills:** The gills are also permeable to respiratory gases, ammonia waste products, and ions. Therefore, while water moves towards the higher osmotic pressure of the blood, sodium and chloride ions also diffuse out of the fish, moving down their concentration gradients in the external environment. To balance this loss, they reabsorb salt from their urine before its excretion and actively uptake salt from water using special cells in the gills (Fig. 2.14).

   Special cells in gill lamellae contain sodium and chloride "pumps". These pumps are special enzymes that use energy to move the ions up their concentration gradient that is from water into body fluid. Besides sodium and chloride, salts which are absorbed by special cells at the base of gill lamellae are Li, Na, Ca, Cl, Br, SO_4 and PO_4. This counteracts the salts loss that occurs through gill.

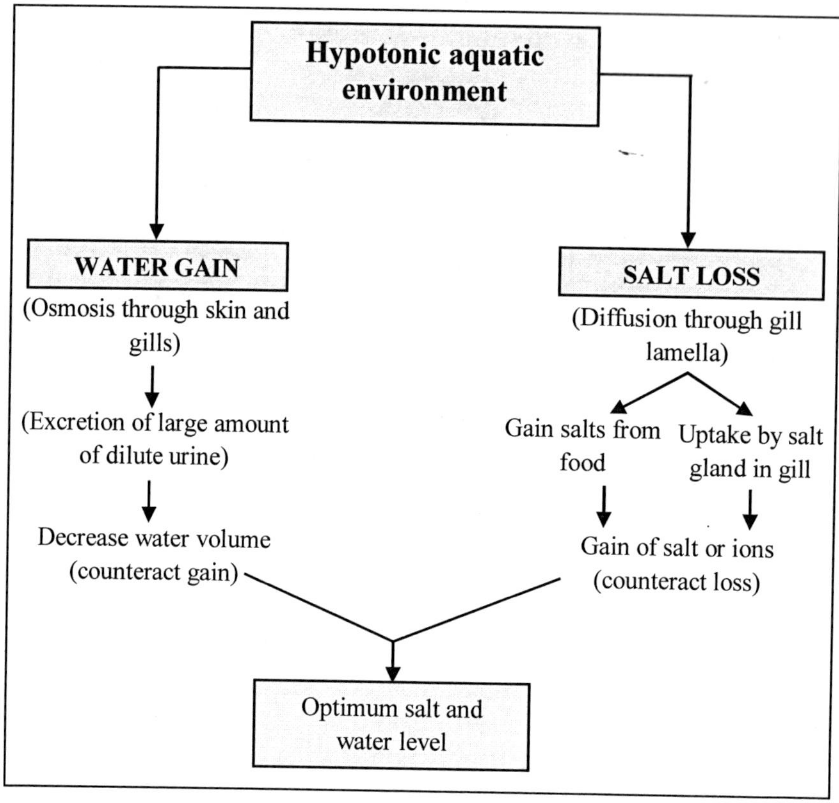

Fig. 2.14: Maintenance of optimum salt and water in freshwater fish through osmoregulation.

2.8.5 Mechanism of osmoregulation in marine water fish

Marine water organisms like fish (example- *Scoliodon*, Blue Angelfish, Black triggerfish) live their lives completely surrounded by water on all sides. It surrounds them externally in their habitat and also comprises much of the body fluid. Fish must, therefore, use some sort of balance between these two separate and very different water environments that is between the body fluid and marine water. This balance is met through the processes of osmoregulation.

Problems: In the marine water, the body fluid of fish has a lower concentration of salt compare to the external water environment. Thus, the body fluid of marine water fish is **hypotonic** to their external watery environment, and thus faces the following problems:

1) As sea water passes over the gill lamella, water diffuse out of the blood (hypertonic) into the sea water by osmosis. Therefore, marine fish must have some mechanism for getting or conserving water.
2) Drinking of sea water increases salt concentration in the blood, and thus they must have a mechanism for removing this excess salts (Fig. 2.15).

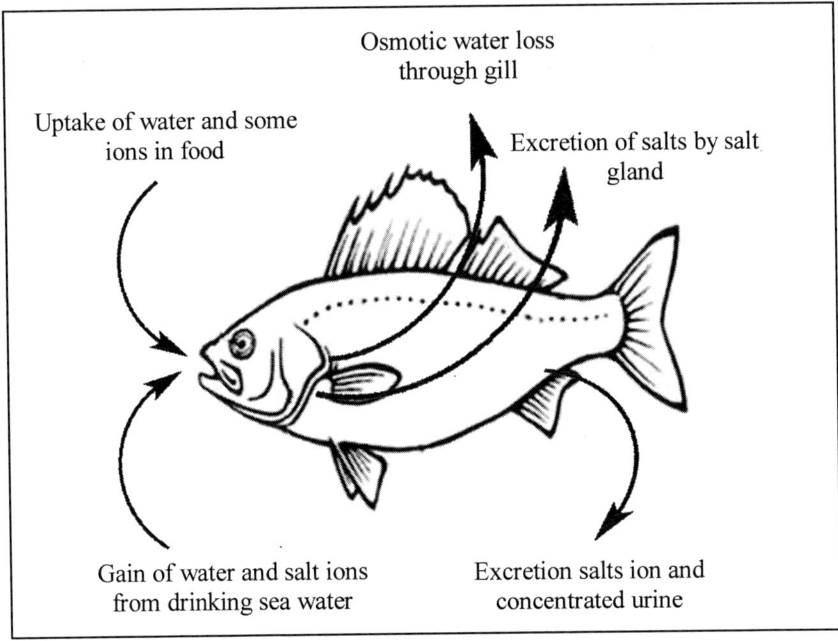

Fig. 2.15: Mechanism of osmoregulation in marine water fishes.

Solution/ osmoregulation- In the marine water fish, kidneys and gills perform the function of osmoregulation in the following ways-

1) **Drinking excess sea water and excretion of less urine:** Marine water fish replace the water loss by drinking a large amount of sea water. They also have a kidney with relatively few small glomeruli, and thus producing less amount of urine.
2) **Excretion of ions through gills:** Salt is removed by chloride secretory cells in the gills, which actively transport salts from the blood into the surrounding water.

Special cells in gill lamellae contain sodium and chloride "pumps". These pumps are special enzymes that use energy to move the ions up their concentration gradient that is from body fluid to sea water. The salt is returned to the sea by active transport at the gills.

In addition to this, the kidneys of marine water fish also remove the excess salt into seawater by excretion of salts along with urine.

3) **Osmoregulation by the Rectal Gland/ salt gland:** Marine elasmobranchs (sharks, rays, and skates) faces the problem of a natural and continuous diffusion of salts into the body from the external sea water. In this organism, the rectal gland functions as a salt secreting organ for excreting excess salts from body fluid (Fig. 2.16).

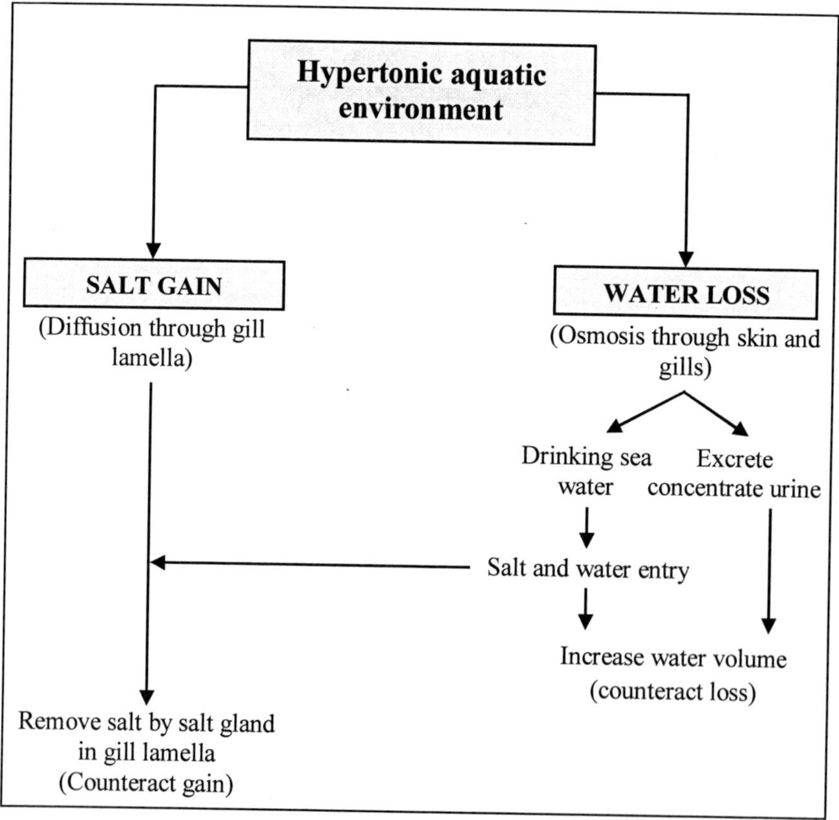

Fig. 2.16: Maintenance of optimum salt and water in marine water fish through osmoregulation.

4) **Osmoregulation by producing metabolic urea:** In contrast, sharks, hagfish and many marine invertebrates use different mechanisms. Their body fluids have almost the same concentration of ions as sea water. The high level of urea, TMAO in marine elasmobranch blood is maintained by the kidneys. Renal tubules are capable of reabsorbing urea, ensuring that this important osmoregulatory compound is not wasted. It has been found that bull sharks taken from marine waters possess a mean serum urea level of 356 mM/l and TMAO levels of 46.6 mM/l (Thorson 1973).

2.9 BIOELECTROGENESIS IN FISHES

The generation of electricity by living organisms is called bioelectrogenesis (bio + electro + genesis) and the branch of science that deals with the electrical properties of living cells or tissues is called electrophysiology. Fish exhibiting bioelectrogenesis called electrogenic fish are often electroreceptive that is they also have the abilities to detect electric fields. The output of the electric organ is fish is popularly called electric organ discharge (EOD).

Types of electric fish: On the basis of strength of EOD generated, fishes can be broadly categorised into following two groups:

1) **Strong electric fish:** Fishes under this group are capable of generating enough current to stun prey. Based on surroundings (fresh or salt water), the amplitude of EOD in these fishes ranges from 10 to 600 Volts.

   Examples- electric eel, electric catfishes, electric rays, etc.

2) **Weak electric fish:** As the name suggests, fishes belonging to this group generate low amplitude EOD, generally less than one volt. They mainly use this weak EOD for navigation, object detection and communication.

   Example- Peters' elephantnose fish, black ghost knifefish, etc. In addition to this there are some fishes (e.g., rays, skate, catfish, paddlefish, etc.) called electroreceptive fish can only sense EOD but cannot produce EOD.

Electric organ, its structure and physiology: The electric organ that generates electrical fields around the electric fish is composed of modified skeletal muscle cells. In most of the electric fishes, this organ is located in the tail region. Electric organs of electric fishes like electric eels, rays is composed of several thousand of flat disk-like cells called electrocytes lying one above the other. Each of these cells is capable of producing around 0.15 V of EOD. The physiology of these cells involves pumping of positively charged sodium and potassium ions via transport proteins in the presence of ATP (Fig. 2.17, 2.18).

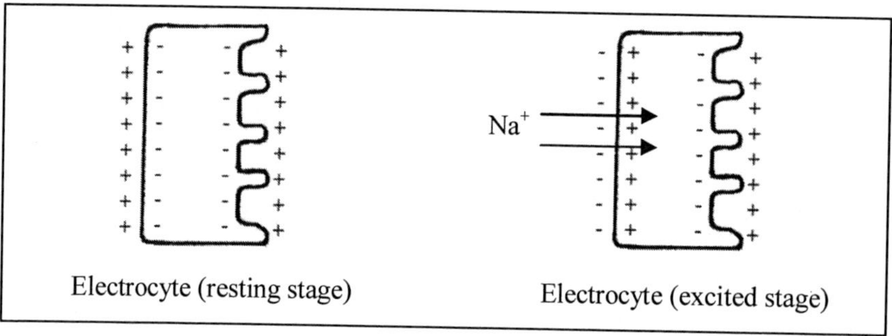

Fig. 2.17: Resting and excited electrocyte.

At rest, the interior of each electrocyte, remain negatively charged with respect to the external surfaces. When a nerve impulse reaches the posterior surface, Na^+/K^+-ATPase pump activated and the rapid inflow of sodium ions momentarily reverses the charge. As a result, the posterior surface of each electrocytes has now become negative, while the anterior surface remains positive. The charges now strengthen each other and current flows similar to an electric battery with the cells wired in "series".

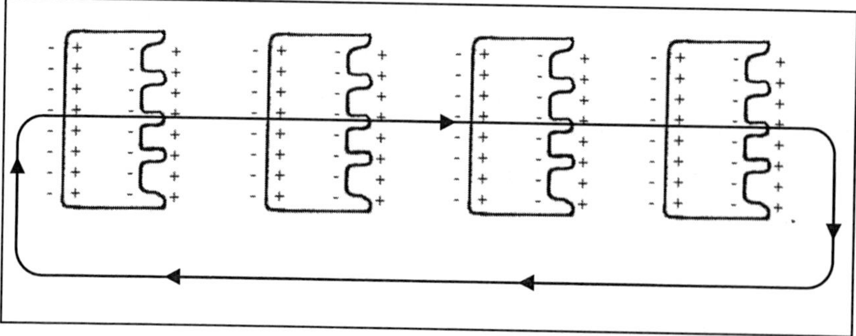

Fig. 2.18: Mechanism of production of high electric organ discharge (EOD) by electrocytes.

Pacemaker nucleus found in the bodies of electric fish is the main regulator of the entire process. For example, when an electric eel detects its prey, the pacemaker neurons are stimulated and the signal is transmitted to electrocytes of the electric organ by neurotransmitter acetylcholine. This results in changes in ion concentration and subsequent generation of EOD.

Biological significance of electric organ: The EOD of electric fishes is mainly utilised for electrolocation, self-defense, electrocommunication and prey stunning.

1) **Predators defence/stunning prey:** Species of electric fish belonging to the strong electric fish group (e.g., strongly electric catfishes, torpedo, and electric eels) are capable of generating extremely strong electric discharges (hundreds of volts) from their electric organs. Such strong discharges are either used to stun the prey or to ward off the predator.
2) **Electrolocation/Navigation:** Weak electric fish typically use their low volt EOD for electrolocation. Electric fish can detect the distortions caused by any objects in their electrical field, and thus can obtain information about the distance of nearby objects or organisms. The electric eel also uses low voltage EOD for navigation in turbid habitat.
3) **Communication:** Electric fishes often communicate with each other by sensing the EODs of each other. Theses discharges are often species specific and sex specific. For example, the males of the nocturnal *Brachyhypopomus pinnicaudatus*, native to the Amazon basin, give off big, long electric call to attract a mate.
4) **Self defence by mimicry:** Some species of American river fish commonly known as bluntnose knifefish generate an electric discharge that is totally similar to the EOD of the electric eel. This is hypothesised that it is a Batesian mimicry of the dangerous eel (Stoddard 1999).

2.10 BIOLUMINESCENCE

Bioluminescence is a form of chemiluminescence and it refers to the production and emission of light by living organisms through a chemical reaction. Bioluminescent creatures occur throughout the marine environment from the ocean surface to the ocean floor.

Example- Bioluminescence occurs in diverse groups of organism in the animal kingdom, including marine vertebrates and invertebrates, as well as in some fungi, microorganisms (bioluminescent bacteria) and terrestrial invertebrates such as fireflies.

Chemistry of Bioluminescence: As it is a form of chemiluminescence, it involves a chemical reaction for emission of light. Following are the different types of chemical reactions generally observed in bioluminescent organism:

1) **Luciferase-based bioluminescence:** In nature, bioluminescence produces different colours, mainly blue, green and yellow. The distinctive colour of light that a species emits depends on the environment in which it has evolved. The principal chemical reaction in bioluminescence involves the light-emitting pigment luciferin and the enzyme luciferase. It involves-
   - Oxidation of luciferin: Luciferin reacts with oxygen and converted into oxyluciferin.
   - Cofactors: In some species, the type of luciferin requires cofactors such as calcium or magnesium ions and the energy-carrying molecule ATP.
   - Byproducts: Carbon dioxide (CO2), adenosine monophosphate (AMP) and phosphate groups (PPi) are released as waste products
   - Light intensity: Each conversion of luciferin to oxyluciferin, generates around 560nm of light.

$$\mathbf{L + O_2 + ATP \xrightarrow[Mg^{2+}]{Luciferase} oxy{-}L + CO_2 + AMP + PP + lightenergy}$$

2) **Aequorin system:** The parchment worm Chaetopterus (a marine Polychaete) use photoprotein, aequorin, instead of luciferase to generate light. Aequorin is a calcium-activated photoprotein. It is quite different from luciferase as it does not require oxygen and other cofactors. The reaction steps involve:
   - The photoprotein consists of two components: an apoprotein (apoaequorin) and a chromophore. The chromophore is made up of coelenterazine and molecular oxygen
   - When Ca^{2+} binds to aequorin, an intramolecular reaction takes place in which coelenterazine is oxidised to coelenteramide.

This yields light (~470 nm), $C0_2$, and a blue fluorescent protein. The emission of light is mainly due to the decay of the bound coelenteramide from an excited state.

Apoaequorin-chromophore + nCa^{++} ⟶ nCa^{++}-protein-oxyluciferin complex + CO_2 + hi (469 nm)

3) **Modified aequorin system:** This is found in hydrozoan jellyfish, *Aequorea victoria*. In this organism, the blue light released by aequorin in presence calcium ions is absorbed by green fluorescent protein (GFP). The GFP then re-emits it at a longer wavelength as the green light.
4) **Flavin-luciferase system:** Bacterial luciferase catalyses the reaction of reduced flavin mononucleotide ($FMNH_2$) with O_2 to form a 4a-peroxyflavin derivative. The derivative then reacts with a long chain aldehyde and emits blue-green light (490nm) along with the formation of FMN, H_2O, and the corresponding fatty acid.

Biological significance of bioluminescence: Among the fishes, almost all bioluminescent species are marine animals. The uses of bioluminescence by animals include counter-illumination camouflage, mimicry of other animals, defence, detection of prey, and signalling to other individuals of the same species, such as to attract mates.

A classic example, where a single fish use bioluminescence for more than one purpose is blackdragon fish (*Melanostomias bartonbeani*). The fish has a flashlight near to each eye that it can "turn on" to look for prey or signal mates. It has a light emitting barbel which is used as a lure to attract the prey. It has light organs arrayed along its belly that will hide its shape by counterillumination. Finally, it has a luminous body that makes the fish a good burglar alarm displayer.

1) **Counterillumination camouflage:** In many animals of the deep sea, bioluminescence is used for camouflage by counterillumination in which an animal produces light to match an illuminated background. For example, hatchetfishes (family Sternoptychidae) lanternfish or lampfish (family Myctophidae) use their photophores on their undersides, heads and tails as camouflage by producing a counter-illumination. Light emitted from their undersides exactly match the color and intensity of sunlight penetrating the water from above,

making them invisible from below. Likewise, at night, their dark unlighted backs merge with the darkness of the deep ocean water and they remain invisible from above, keeping them safe from seabirds.

2) **Attraction:** Many fishes make symbiotic relationships with bioluminescent bacteria to use their light as a lure to attract the prey. For example, species of deep-sea anglerfishes, flashlight fishes, footballfish (family Himantolophidae) use symbiotic bacteria to produce light. Deep-sea anglerfish use their symbiotic bacteria to emit light from their esca mainly to attract prey. On the other hand, viperfish like a fishing lure by flashing the light of photophores on and off attract smaller fish.

3) **Puzzling predators:** Use of bioluminescence for defence against predators is common among deep-sea animals. They shine when they detect a predator, possibly either to confuse the predator or making the predator itself more vulnerable by attracting the attention of predators from higher trophic levels.

   For instance, the black dragonfish (*Idiacanthus atlanticus*) has photocytes in their fins and along the tail. If the drangonfish is disturbed by a predator, these photocytes flash brightly to frighten the predators.

4) **Mimicry:** Bioluminescence is used by a number of predator fish to mimic other harmless species called aggressive mimicry. Many species of deep sea fish such as the anglerfish and dragonfish use aggressive mimicry to attract the prey.

5) **Detection of prey:** Some deep sea barbeled dragonfishes emit a red glow. This adaptation allows the fish to see red-pigmented prey, which is normally invisible in the deep ocean environment.

6) **Communication:** The photophores of lanternfishes (family Myctophidae) produce flashes of light either to illuminate the prey or to get the prey to fluoresce in response or to surprise the predators. The arrangements photophores in lanternfish are species-specific, and thus it is thought that in these fish bioluminescence play the key role in communication, particularly in shoaling and courtship behaviour.

7) **Reproduction:** Bioluminescence can be used for sustaining aggregation and luring mates in the deep sea (Rivers and Morin 2012). Some fish (lanternfish) species have species and sex specific luminescent signal. For example, deep-sea sharks have species identification bands and females have a bright line down their

pectoral fin in the mating season not only to attract mates but also to guide the males where to clamp onto during mating.

8) **Burglar alarm display:** When grabbed or threatened by a predator, fishes like blackdragon fish start emitting light either to scare the predator to go away, or to attract the attention of a larger predator that may attack their attacker. This type of flashy bioluminescent display is popularly called **burglar alarm** because like an alarm, it also attracts attention for help.

Application in research and decoration:

1) **Research:** In the molecular biological laboratory, luciferase-based systems are commonly used in genetic engineering as a reporter gene to test the strength or regulation of a regulatory sequence.
2) **Decoration:** Many companies are trying to use bioluminescent lights for shop fronts, street signs and decorative lighting.

2.11 SHOALING AND SCHOOLING

These are the two common, prominent social and group behaviour exhibited by many of the fish species. When a group of fish stays closely together, it is called shoaling. It is very common both in the fresh and marine environment. Example- The common shoaling fishes are pygmy corydoras or pygmy catfish, cardinal tetra, rummy nose tetra, ruby tetra, etc.

On the other hand, schooling refers to the swimming of a group of fishes of a species in a particular direction in a synchronised or coordinated manner. It is very common in the marine environment. The fish performing schooling behaviour can change directions and turn as a whole. Schooling fish are commonly found in the marine habitat. Example- Some of the schooling fishes are Yellowtail fusiliers, Tuna, blacksmith and Herring.

Characteristics of schooling: The important characteristics of fishes in a school are as follows:

1) A school is made up of members of a single species of very similar size.
2) Fishes in school mimic and follow each other's movements and always keep a defined spacing among them.

3) Fishes in a school use two senses to maintain the school and are visual cues and their lateral line systems.
4) There is no particular leader in a school rather all the members in a school mutually attract to each other (Maier 1998).
5) Members in a school remain close to neighbours and always avoid collisions with their neighbours.

Significance of schooling: Fish species capable of performing schooling behaviour offer numerous benefits to individual fish including foraging advantage, reproductive advantage, and increased protection from predators. Some of these are briefly discussed here:

1) **Hydrodynamic advantage:** Due to regular spacing and size uniformity, a fish in schools obtain a hydrodynamic advantage, and thus reduces the cost of swimming.
2) **Predator avoidance:** The primary advantage of forming schools is to minimise the chances of being eaten by larger fish. Predator faces this difficulty as the schooling behaviour can confuse the lateral line organ and the electrosensory system of predators (Larsson 2012). In addition to this, many moving targets in a school create a sensory overload on the visual activity of the predator popularly called **confusion** effect. Thus, simply being a part of a school reduces the chances of being killed is popularly called **numerical dilution effect**.
3) **Advantages in food searching:** It has been found that swimming in groups enhances foraging success rate. The simple reason behind the success is the presence of many eyes of a school searching for the food. Schooling herrings, Indian mackerel employ a method called **ram feeding** (fish moves forward, keeping its mouth open, and thus engulfing the target prey along with the water) to feed on very alert and evasive copepods.
4) **Advantages in reproduction:** Both shoaling and schooling behaviour has reproductive advantages as finding a potential mate in a school does not take much energy.
5) **Advantages in migration:** Navigation performed by a large group fishes in a school will be always better than navigation performed by an individual fish. This is the reason why fishes showing, schooling behaviour can perform long distance migrations. For example, some herrings have their spawning ground in southern Norway, their nursery ground in northern Norway and their feeding ground in Iceland.

6) **Increased vigilance:** This is also an anti-predator effect which can be explained by many eyes hypothesis. A group fishes always have better capacity of scanning the surrounding environment for predators than a single fish.

2.12 PARENTAL CARE IN FISH

2.12.1 Parental care

Parental care refers to any behavioural strategy where parent invests time, energy, etc. for increased survivorship or evolutionary fitness of their offspring. Caring may be performed by the mother alone (maternal care) or the father alone (paternal care) or both the parents (biparental care).

The basic theory behind parental care is to spend more effort on a relatively small number of offspring for increasing the chance of their survival rather produces a large number of offspring. Parental care is common throughout the animal kingdom starting from invertebrates to mammals. The two common means of parental caring include feeding and protecting its offspring from the adverse environment or from predators.

2.12.2 Patterns and diversity of parental care in fishes

Though parental care is less common in fishes due to their capacity to produce a large number of eggs and sperms, a number of fish species (~25%) have evolved parental care, mainly paternal. Fish parental care ranges from hiding of eggs, protection of fry, cleaning, fanning of eggs, carrying young on the parent's body to the feeding of young. Some of the notable examples parental care found in fishes are highlighted below:

1) **Mouthbrooding or buccal incubation:** Mouthbrooder fishes hold their eggs or fry in their mouth for extended periods of time until the eggs hatch or fry become a free swimmer. Fish species of African and South American cichlids (family Cichlidae), black-chin tilapia show maternal mouthbrooding; arowana (family Osteoglossidae), betta and sea catfish show paternal mouthbrooding, while genus of cichlid *Xenotilapia* and spatula-barbled catfish show bipaternal mouthbrooding.

In most of the paternal care, after courtship, the male fertilises the eggs and collects and holding them in his mouth until they hatch. However, in cichlids (maternal caring), the female first collects the eggs into her mouth and then male fertilise the eggs in situ.

2) **Male pregnancy:** Male pregnancy refers to caring of brood by the male. Unlike most animal species where the offspring is usually carried by the female partner, in fishes belonging to the family Syngnathidae (seahorses, pipefish, weedy and leafy seadragons), males carry out this function. The males of these species possess a brood pouch on their trunk or tail, where the female lays eggs. Males help in the development of embryo by a) maintaining a proper pH in the pouch for the development of embryos and b) supplying nutrients (glucose and amino acids) to the embryos through the highly vascular wall of brood pouch.

3) **Hiding of eggs in nests:** The simplest form of parental care found in fish is the hiding of eggs inside the nests after egg laying. For example, the female of salmon and rainbow trout build nest called **redds** (1 to 3 feet in depth) by digging in the gravel of streams or the shoreline of lakes using their tails. The eggs that are laid in these redds are fertilised and masked by the parent till they are hatched.

In contrast to this simple type of nest, males of some fish species are an amazing nest builder. For example, Lake Malawi cichlid (*Cytocara eucinostomus*) and stickleback (*Gasterosteus aculeatus*) build nests by glueing together dead aquatic plants using a sticky glycoprotein known as spiggin, secreted from their kidneys. The male then attracts the females by courtship display to lay eggs in the nests. After fertilisation, male continue his service of protecting the eggs from intruders and predators until they hatch. During this incubation period, the male also oxygenates the eggs and removes rotten eggs and debris from the nest.

Many species of fish also protect their eggs by forming bubble nests or foam nests. The bubble nest is floating masses of bubbles blown with an oral secretion and the fish that are capable of building and guarding bubble nests are known as **aphrophils**. The common examples of bubble nest makers or aphrophils are gouramis, *Betta* species, synbranchid eel (*Monopterus alba*), *Ctenopoma* (Family- Anabantidae), *Polycentropsis* (Family- Nandidae), *Hepsetus odoe* (Family- Hepsetidae), and electric eel etc. The duration of time spent protecting young ranges from 1 day in the Sacramento perch, (*Archoplites interruptus*) to over 4 months in the Antarctic plunderfish (*Harpagifer bipinis*) (Balshine and Sloman 2011).

4) **Mermaid's purse or devil's purse:** Some egg laying sharks (bullhead sharks, carpet sharks, catsharks, and zebra shark), skates, and chimaeras lay their eggs inside an egg capsule called **Mermaid's purse** made up of collagen protein. These egg cases often possess adhesive fibers or long tendrils that serve to anchor them to seaweeds growing on the seabed. After hatching young fishes come out of the Mermaid's purse by rupturing its wall.

5) **Fanning and cleaning of eggs:** Many fishes like stickleback and bluegills regularly aerating their eggs using their pelvic or pectoral fins. By fanning the eggs parents allows fresh and aerated water to pass over the eggs. This fanning behaviour is highly developed in American lungfish *Lepidosiren*.

   During spawning season the breeding males of American lungfish developed highly vascularized, gill-like pelvic fins through which oxygen diffuses from the bloodstream into the poorly oxygenated water surrounding the eggs or young. Many fish use their mouth for sucking the detritus from healthy eggs and dead or fungal infected eggs.

6) **Feeding of young by trophoic eggs:** Perhaps the most complex or highly structured, but less common form of parental care found in fishes is feeding their young or fry. For instance, females of Kampoyo catfish (*Bagrus meridionalis*) do not lay all the eggs during breeding time rather she keeps some unfertilised eggs within her ovaries. When the fry are 15 days old, mothers force out these unfertilised eggs popularly called **trophic eggs** and feed her fry with these trophic eggs. Every day, mostly in the morning, young line up at her vent and feed on the small eggs she exudes (McKaye 1986). In addition to maternal care, following types of paternal caring behaviours are also found in fishes:

   a) The father often ploughs into the ground and stir up debris on which the young feed.
   b) The father often takes sediments into his mouth and releases it near the nest after churning it inside his mouth.
   c) The young also often feed on the invertebrates (planktons) found on the gills of the male parent.

In many cichlids such as discus fish, Midas cichlid, angelfish and orange chromide both parents secret mucous from mucus-producing glands found in the skin. Up to 2 weeks, fry feed on this nutrient rich skin mucus. Parents of several cichlids species belonging to the genus *Cichlasoma* often stir up the gravel by fin digging to feed their young with this stirred-up material.

7) **Signaling and retrieving fry at the time of danger:** Signaling of danger to the young, a form of parental care is found in many species of fishes. Common types of danger signals found in fishes include flickering of the pelvic fins up and down (cichlid), brief jerks of the head or twitches of the whole body and slow backwards swimming with head down. The parental male of Siamese fighting fish when senses danger; he generates wavelets by shaking his pectoral fins. In response, the young swim in the direction of the wave source and enter into the buccal cavity of the male parent. The male brings them to a safe place where it spits them out (Kang and Lee 2010).

8) **Broken wing display:** Broken wing display or distraction display is a form of nest protection behaviour commonly found in ground-nesting birds where they lure predators away from their nest. A similar type of behaviour is found in three-spined sticklebacks. Shoal of female three-spined sticklebacks often fall upon the nest of a parental male and eat all the eggs (cannibalism), called **nest-raiding**. In a defensive behaviour, when a male fish sees a shoal of females coming towards the nest, he swims a short distance away from his nest and starts poking his snout into the ground. In this way male copy the nest-raiding behaviour of female and saves his nest by fooling the shoal that a nest has been identified there. The confused shoal then rushes to this site and start digging there.

9) **Egg dumping or brood parasitism:** Egg dumping refers to the behaviour where females deposit her eggs inside the nest of other parents and let them take care of the eggs. For example, some species of minnows spawn in the nest of sunfishes, while golden shiners (*Notemigonus crysoleucas*) choose the nest of two of its predators, the bowfin and the largemouth bass for egg dumping. The female catfish *Synodontis multipunctatus* lay eggs at the same time and in the same spot, where various mouthbrooding cichlids lay their eggs. By doing this, the catfish uses the mouthbrood of cichlids

for the protection of her eggs. The female cichlid when picks her eggs also picks up the indistinguishable eggs of catfish and incubate them inside her mouthbrood until the egg hatches (Sato 1986).

10) **Mobbing:** It is a form of antipredator behaviour where individuals of a species form a mob and frighten the predator by cooperatively attacking or harassing it. Some species of fish use mobbing to protect their offspring. For example, nesting males, gravid females of bluegill sunfish form a mob when they find snapping turtles approaching towards the spawning colony. The mob performs threat displays without physically attacking the snapping turtles until it left the colony area.

11) **Viviparity and ovoviviparity:** The highest degree of parental care in the form of viviparity is rare in bony fishes, although it has evolved in several cartilaginous fishes. Fertilisation is internal and the embryo is nourished by the mother's body. Examples include blue shark, bull shark, rose fish, dogfish (*Scoliodon*), lemon shark, four-eyed fish, etc. Recently, it has been discovered that pregnant female European eelpout (*Zoarces viviparus*) suckles young embryos. This species has a long gestation period of approximately six months. After depleting the egg's yolk reserves, the young eelpouts attach their mouths to ovarian follicle of their mother and suckle fluid is rich in proteins, fatty acids, glucose and oxygen (Skov 2010).

    In ovoviviparous fishes fertilisation is also internal, but unlike viviparous, their embryo is nourished by egg yolk. Tiger shark, bat ray, blue stingray, coelacanth, etc. are the common examples of ovoviviparous fishes.

Costs and benefits of parental care: Substantial weight loss of parents is the common cost associated with parental care. Weight loss is primarily due to a) predator chasing b) lack of foraging time c) inability to take in food, for example, mouthbrooders cannot eat food during buccal incubation. On the other hand, for better reproductive success parents spent more effort on a relatively small number of offspring for increasing the chance of their survival rather produce a large number of offspring. Parents improve the chance of offspring survival or fitness by protecting descendants from predators (building nests), maintain an environment that favour their development (fanning, cleaning), feeding young, etc.

Hormonal regulation of parental care: A very few hormones and neuropeptides have been found to be associated with parental care regulation in fish species. Positive correlation between high plasma androgen level (testosterone and 11-ketotestosterone) and different parental care behaviours like nest building, competing for territories is found in many fishes. In addition to this, prolactin is found to stimulate fanning behaviour in the three-spined stickleback and in bluegill (Balshine and Sloman 2011).

2.13 MIGRATION IN FISHES

2.13.1 Migration

Migration refers to the physical movement of animals (including fishes) from one place to another on a regular basis. Some fishes migrate on a regular basis, while some on time scales ranging from daily to annually or longer. The migratory distance ranges from a few metres to thousands of kilometres.

If the migration of the fish occurs with the flow of water current, it is called as denatant movement. On the other hand, if the direction of migration is against the water current, it is known as contrary movement.

2.13.2 Types of migration in fishes

Fishes live either in freshwater habitat or in the marine habitat. On the basis of their direction of migration or environment of the migratory area, migration in fishes can be classified into the following types:

1) **Latitudinal migration:** Migration of fishes from north to south and back again is called latitudinal migration.
   Example- Barracudas, Swordfish, etc.
2) **Potamodromous migration or potamodromy:** In potamodromous migration, fishes migrate from one freshwater habitat to another in search of food or for spawning. Thus, a fish potamodromous spends its whole life in fresh water. This type of migration covers short distances generally from an upstream tributary to a mainstream river or between connected lake and river systems.

Example- Lake sturgeon, corps, Common Asiatic carps, Trouts, Mahseer, etc.

3) **Oceanodromous migration or oceanodrmy:** In this type of migration fishes migrate within the sea water to take the advantage of favourable conditions wherever they occur.

   Example- Herrings, Sardines, Mackerels, Cods, Roaches and Tunas migrate to area favourable for spawning and then return to return to their parental area.

4) **Diadromous migration or diadromy:** Migration in which fishes migrate from fresh water to sea or from sea to fresh water is called diadromous migration. It can be further classified into the following two types

   a) **Catadromous migration (fresh to sea):** Migration in which fishes migrate from fresh water down into the sea water to spawn is called catadromous migration (Fig. 2.19).

      In catadromous migration, fish born in salt water, then migrate into freshwater as juveniles where they grow into adults before migrating back into the sea to spawn. Example- American eel, European eel, Inanga, Shortfin eel, Longfin eel, etc.

      The most remarkable catadromous fishes are the two species, freshwater eels namely European eel (*Anguilla vulgaris*) and the American eel (*Anguilla rostrata*). The adult eels that spend 8-15 years in continental rivers are about a metre long and yellow in colour. Before migration and with the advent of autumn their a) body colour changes from yellow to metallic silvery grey, b) digestive tract shrinks and they stop feeding c) gonads get fully matured and enlarged, d) eye enlarges, vision sharpens and sensory organs become more sensitive and e) they deposit large amount of fat in their bodies as reserve food. They then migrate to the Sargasso Sea off Bermuda in South Atlantic by crossing the Atlantic Ocean and covering a distance of approximately 5,600 km during the journey. Males joined the females in coastal areas of the sea and then migrate togetherly to the Sargasso Sea. Each female eel lays about 20 million eggs and are soon fertilised by males. Immediately after spawning in the Sargasso Sea off Bermuda, the adults die.

      Fertilised eggs hatch into leaf-like, semitransparent, pelagic larvae called **Leptocephalus**. These larvae then start their journey towards the continental rivers (parent's home) and upon reaching coastal waters they metamorphose into cylindrical miniature transparent eels called **Elver** or **Glass eel**. Female elvers continue their journey and ascend to the continental rivers and metamorphose into yellow-coloured adults, while males remain behind on the coast and wait there for the females to return for spawning journey towards the Sargasso Sea.

b) **Anadromous migration (sea to fresh):** Migration in which fishes migrate from sea water up into the fresh water for spawning is called anadromous migration. In anadromous migration, fish born in fresh water, then migrate to the sea as juveniles where they grow into adults before migrating back into the freshwater to spawn. Example- Atlantic salmon (*Salmo salar*), Pacific salmon (*Oncorhynchus spp.*), three-spined stickleback, shad, etc.

The best known, anadromous fish is Atlantic salmon. Atlantic salmon spend between one to four years at the marine environment. Adult salmons living in the sea are metallic silvery grey in colour, but before migration their body colour changes to reddish-brown, they grow a hump, develop canine-like teeth and develop a **kype** (a hook-like secondary sex characteristic at the lower jaw of male salmon). During winter, they enter rivers and start moving hundreds of miles upstream against the water currents (contranatent) to reach mountain streams. In the mountain streams, the female salmon build many spawning nests called **redd** by digging in the gravel of streams with her caudal fin. Within these spawning nests, female lays eggs and the male releases milt over them.

In most cases, both the adults die within a few days or weeks of spawning a trait called **semelparity**. However, 2 and 4% of Atlantic salmon survive, return to sea and spawn again a trait called **Iteroparity** (capable of spawning more than once in life). After spawning, the eggs hatch within 2-3 months in the following spring. The juvenile stage of the salmon called **Alvin** remains within the nest and obtains its nourishment from the yolk sac attached to its belly. Alvin subsequently transforms into fry which feeds on planktons and grows into fingerlings. They start swimming along with the water current and change into **Smolt** (young salmon covered with silvery scales). The Smolts gather at the river mouth in large numbers and enter the sea to metamorphose into adult salmons.

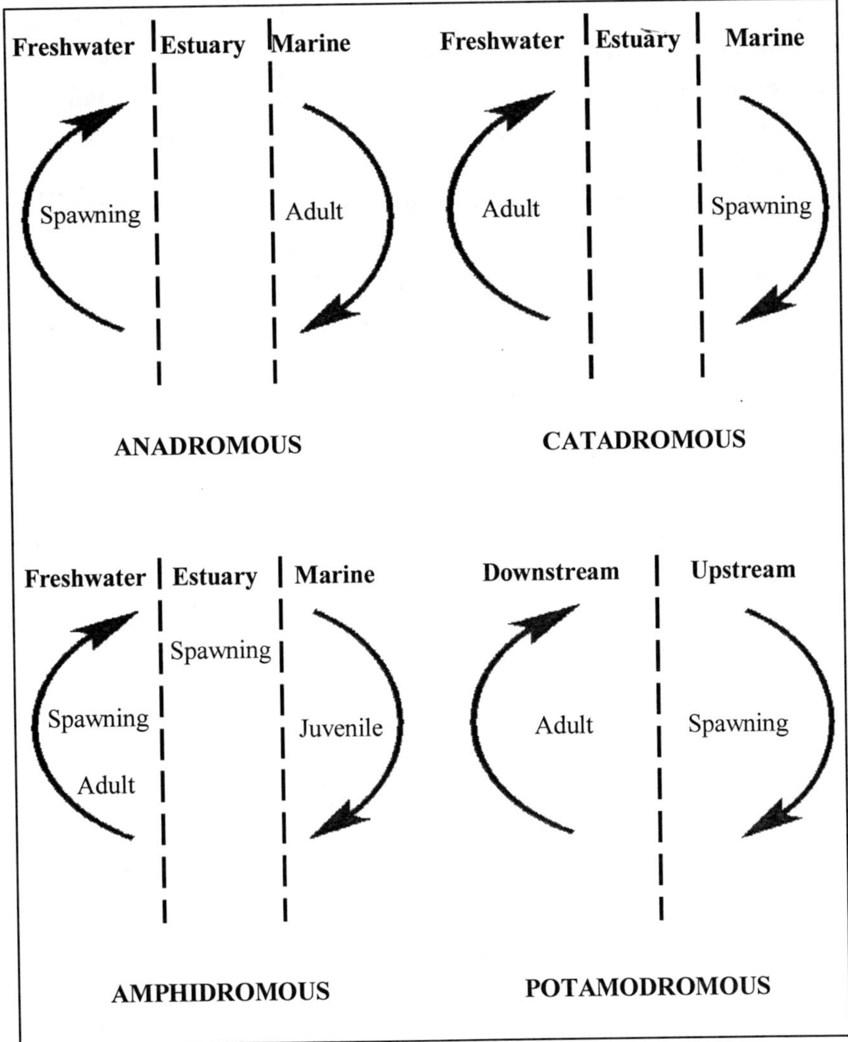

Fig. 2.19: Different types fish migration.

c) **Amphidromous migration:** Migration in which a fish migrate from fresh water to the sea water or vice versa, not for the purpose of breeding is called amphidromous migration. Example- Bigmouth sleeper, Mountain mullet, River goby, Torrent fish, etc. The amphidromous fish migrate regularly at some particular stage of the life cycle.

2.13.3 Causes of fish migration

Animals including fishes migrate from one place to another for feeding, breeding or to escape adverse weather extremes. Some of the reasons or importances behind the migration in fishes are briefly discussed below:

1) **Feeding or alimental migration:** If the migration is mainly in search of food, then it is called alimental migration. Fishes with large population size like salmons, cods and sword fish exhaust food resources in an area very quickly. These fishes migrate constantly in the open sea in search of food. Thus, due to population pressure fish migrate in search of new favourable areas where food is abundant and competition is minimum.
2) **Spawning or reproductive migration:** If the migration is for breeding, then it is called gametic migration. Many fishes have spawning grounds far away from feeding places, and thus these fishes often migrate between feeding and breeding places. Migration in both the catadromous (example- eel) and anadromous (example- salmon) fishes is mainly for breeding.
3) **Juvenile migration:** In this type of migration, larval stages of fishes which hatch in spawning grounds migrate long distances in order to reach the feeding habitats of their parents. For example, anadromous fish born in freshwater and then migrate to the sea as juveniles where they grow into adults.
4) **Seasonal migration:** It occurs mainly due to the adverse surrounding environment. For example, in summer climatic condition is favourable and the food is abundant in arctic areas, but during winter season temperature fall below zero and food becomes scarce. Thus, fishes that inhabit arctic areas must migrate towards the subtropical and tropical areas to escape the extremes of weather conditions. Similarly, many fish migrate from lakes out into the neighbouring streams in the winter.
5) **Protective migration:** Migration that occurs due to spawning, feeding and wintering are also called protective migration as this behaviour ensure a further life of fish in their environment.

2.13.4 Factors influencing migration

Physical, chemical and biological factors influence the migrations in fishes. Physical factors like depth of water, temperature, light penetration, photoperiod, turbidity, the velocity of current influence fishes to undertake migration. Chemical factors are pH and salinity. Biological factors are hormonal action, the maturity of gonads, food, blood pressure, etc. For example, diadromous fishes migrate between fresh and marine waters in their entire life cycle for spawning. The presence of predators (predation pressure) in the area also influences the migration of fishes. For example, many believed that winter migration in fishes from lakes out into the neighbouring streams has been developed as an adaptation against predation. The salinity of water plays an important role in fish migration. Species which are euryhaline like salmon, hilsa are good migrator (long distance migration) as they can tolerate a wide range of salinity. Solar or lunar cycles also affect the migration in fishes. Fishes like herrings generally migrate during the full moon, while sturgeon fishes migrate in the night time. In addition to this pollutant and disturbances in the habitat may force the fish to migrate.

2.13.5 Advantages of migration

Migration in fishes requires the use of a lot of energy, and thus it must be advantageous to the species. Following are some of the advantages of migration:

1) Migration is considered to be an adaptation to high competition between adult and fingerlings for food, and thus migration ensures reproductive success of the group. The spawning grounds may not have enough food and therefore both mature and young individuals of the large population cannot be maintained there. Thus, fish with separate spawning and feeding habitat is advantageous.
2) During spawning, many different populations often meet at the "breeding grounds". Thus, migration increases genetic diversity due to breeding among individuals from a different population.
3) Reduces predation pressure. Many believed that winter migration in fishes from lakes out into the neighbouring streams has been developed as an adaptation against predation pressure.
4) Animals remain in favourable conditions by avoiding extremes of weather conditions like extreme low temperature.
5) Migration enables fishes to explore new favourable areas in the open sea, where food is abundant and competition is minimum.

2.14 REFERENCES

Bainbridge R (1958) The speed of swimming of fish as related to size and to the frequency and amplitude of the tail beat. Journal of Experimental Biology. 35: 109-133.

Baldauf SA, Bakker TCM, Herder F, Kullmann H, Thunken T (2010) Male mate choice scales female ornament allometry in a cichlid fish. BMC Evolutionary Biology. 10: 301.

Balshine S and Sloman KA (2011) Parental Care in Fishes. In: Farrell AP, ed. Encyclopedia of Fish Physiology: From Genome to Environment. San Diego: Academic Press. Vol 1, pp. 670–677.

Brand RA (2008) Origin and Comparative Anatomy of the Pectoral Limb. Clinical Orthopaedics and Related Research. 466 (3): 531–542.

Buckland-Nicks JA, Gillis M, Reimchen TE (2011) Neural network detected in a presumed vestigial trait: ultrastructure of the salmonid adipose fin. Proceedings of the Royal Society B: Biological Sciences. 279(1728): 553–563.

Dehadrai PV (1962) Respiratory function of the swimbladder of *Notopterus* (Lacepede). Journal of Zoology. 139(2): 341–357.

Fish F and Hui CA (1991) Dolphin swimming—a review. Mammal Reviews. 21(4):181–95.

Goodrich ES (1906). Memoirs: Notes on the development, structure, and origin of the median and paired fins of fish. Journal of Cell Science. 50 (198): 333–376.

Hammerschlag N (2006) Osmoregulation in elasmobranchs: A review for fish biologists, behaviourists and ecologists. Marine and Freshwater Behaviour and Physiology. 39(3): 209-228.

Hirata K and Kawai S (2001) Hydrodynamic performance of stream-lined body. National Maritime Research Institute, Tokyo, available at www. nmri. go. jp/ eng/ khirata/fish/ experiment/ upf2001/ body_ e.html.

Hoar WS and Randall DJ (1978) Ferminology to describe swimming activity in fish, p. xiii-xiv. Fish physiology. Vol. VII. Academic Press, New York, NY.

Johansen, K. (1970). Air breathing in fishes. In: Hoar WS, Randall WS, eds. Fish Physiology, Academic Press, New York. Vol. 4, pp. 361–411.

Kang CK and Lee TH (2010) The pharyngeal organ in the buccal cavity of the male Siamese fighting fish, *Betta splendens*, supplies mucus for building bubble nests. Zoological Science. 27(11): 861-866.

Kutschera U (2005) Predator-driven macroevolution in flyingfishes inferred from behavioural studies: historical controversies and a hypothesis. Annals of the History and Philosophy of Biology. 10: 59–77.

Lagler KF, Bardach JE, Miller RR (1962) Ichthyology, 1st edn. John Wiley & Sons, Inc, Hoboken, New Jersey, pp. 1-545.

Larsson M (2012) Why do fish school? Current Zoology. 58 116-128.

Lauder GV and Drucker EG (2002) Forces, fishes, and fluids: hydrodynamic mechanisms of aquatic locomotion. News in physiological sciences.17: 235-240.

Maier, MW (1998) Architecting principles for systems of systems. System Engineering. 1(4): 267–284.

Martin RA (2005) Conservation of freshwater and euryhaline elasmobranchs: a review. Journal of the Marine Biological Association UK. 85, 1049–1073.

McKaye KR (1986) Trophic eggs and parental foraging for young by the catfish *Bagrus meridionalis* of Lake Malawi, Africa. Oecologia. 69(3): 367-369.

Munshi JSD and Hughes GM (1992) Air-Breathing Fishes of India. Oxford and IBH Publishingg Co., New Delhi.

Nwadukwe FO (1995) Hatchery propagation of five hybrid groups by artificial hybridization of *Clarias gariepinus* and *Heterobranchus longifilis* (Clariidae) using dry powdered carp pituitary hormone. Journal of Aquaculture in the Tropics. 10(1): 1-11.

Rivers TJ and Morin JG (2012) Female ostracods respond to and intercept artificial conspecific male luminescent courtship displays. Behavioral Ecology. 24(4): 877-887.

Rosenberger LJ (2001) Pectoral fin locomotion in batoid fishes: undulation versus oscillation. Journal of Experimental Biology. 204 (2):379-394.

Sato T (1986) A brood parasitic catfish of mouthbrooding cichlid fishes in Lake Tanganyika. Nature. 323(6083): 58-59.

Sherwood GD and Rose GA (2003) Influence of swimming form on otolith $\delta^{13}C$ in marine fish. Marine ecology progress series. 258: 283–289

Skov PV, Steffensen JF, Sorensen JF, Qvortrup K (2010) Embryonic suckling and maternal specializations in the live-bearing teleost *Zoarces viviparus.* Journal of Experimental Marine Biology and Ecology. 395(1–2): 120–127.

Stoddard PK (2002) The evolutionary origins of electric signal complexity. Journal of Physiology - Paris. 96 (5–6): 485–491.

Thorson TB, Cowan CM, Watson DE (1973) Body fluid solutes of juveniles and adults of the euryhaline bull shark *Carcharhinus leucas* from freshwater and saline environments. Physiological Zoology. 46(1): 29-42.

*** ***** ***

Chapter 3

FISHERIES

Soumitra Nath

Department of Biotechnology, G. C. College, Silchar, Assam, India

Chapter highlights:

Fish and fisheries are an integral part of most societies and make important contributions to economic and social health and well-being in many countries and areas. Despite this massive importance and value, or because of these features, the world's fish resources are suffering the combined effects of heavy exploitation. Changes in any of the biological, chemical and physical components of the ecosystem can have impacts on the resource and community. Some of these changes may be beyond human control, such as upwelling enriching some coastal ecosystems or large scale temperature anomalies, but these still need to be considered in the management strategy. Others, such as the destruction of aquatic habitats for development, or the direct impact of fishing are due to human action. Thus, fisheries management should both take into account their impacts on the resource and, in consultation with other relevant agencies and parties, take steps to minimise their impacts on the fishery ecosystem. The need for fisheries management arises to find ways in order to protect fishery resources so that sustainable exploitation is possible. The product of the fishing effort and the combined efficiency of the fishing gear and the fishing vessel as well as the skills of the crew, determines the catching capacity.

3.1 INLAND FISHERIES AND MARINE FISHERIES CAPTIVE AND CULTURE

3.1.1 Fishery

Fisheries found within the inland water body locations are commonly called inland fisheries. Inland fishing has been one of the most ancient practices of mankind. Inland fisheries contain water bodies like lakes, rivers, swamps, marshes, rice fields, coastal lagoons and many others. They are spread over large locations over the continents, and thus serve as the highest means of animal protein producers (Welcomme et al. 2010). Fisheries that are based on the marine ecosystem are called marine fisheries. Close to 90% of the world's fishery catches come from oceans and seas, as opposed to inland waters.

3.1.2 Capture techniques

Capture fishing is one of the most common fishing techniques. It encloses every fishing activity that involves capturing wild fish to shellfish (Gabriel et al. 2005). Capture fishing involves various techniques which may or may not contain gears and tools. The modern industrial methods contain numerous tools, and thus capture fishes with high efficiency. The standard industrial ship for the capturing of fishes is the trawler. There are numerous other capture techniques, some of which are briefly discussed below:

- **Fishing without gear:** Fishing without gears is the simplest form of fishing. It involves catching fish with the hands. This is done both by professionals as well as non-professionals. This technique is followed during low tide in shallow water or in deep water. Sometimes picks and hoes are used which are not regarded as gears.
- **Grappling and wounding gear:** It involves the use of tools like spears for catching fish and use of rakes, clamps or tongs to collect shellfish. Other tools such as bows, guns, blowpipes and other projective devices fall in this category.
- **Stunning:** This involves stunning fishes through things such as chemicals, explosions or electric shocks. On stunning, the fishes become helpless and are captured easily. This technique is not

considered a sound environmental technique, and thus has been outlawed.

- **Line fishing:** In this technique, a fishing lure or bait is placed on a line which is mostly used to attract the fish's attention. Many lures are equipped with one or more hooks that are used to catch fish when they strike the lure. Most lures are attached to the end of a fishing line and have various styles of hooks attached to the body and are designed to elicit a strike resulting in a hookset. Modern commercial lures usually are often used with a fishing rod and fishing reel, but there are some who use a technique where they hold the line in their hands. Handlining or handline fishing is a technique in which a single fishing line is held in the hands. In longline fishing, baited hooks are attached at intervals by means of branch lines called snoods or gangions to the main line. When a lure is used for casting, it is continually cast out and retrieved, the retrieve making the lure swim or produce a popping action. A skilled angler can explore many possible hiding places for fish through lure casting such as under logs and on flats.
- **Trapping:** Trapping is a passive way to catch fish, shellfish, crustaceans (crabs, prawns etc.) and cephalopods (octopus, squid, etc.) and is different from active fishing methods such as dredging and trawling. Traps can vary, from simple structures such as rock corrals able to hold various fish species passing by, to highly specialised equipment such as lobster pots.
- **Fishing with bag nets:** Bag nets are containers made of nets that are dragged below the water surface and catch fish. The net is held open by the frame and water current.

3.1.3 Culture techniques

Aquaculture is a form of agriculture that includes the cultivation, propagation and marketing of aquatic animals. Aquaculture is growing up to become an important source of protein for the world's growing population. This is because the capture fish industry has peaked and is likely to decline due to the depletion of the wild stocks. Aquaculture will become an important source of seafood products (Stottrup and McEvoy

2003; Quinitio and Parado-Estepa 2003; Tucker and Hargreaves 2004). The production techniques for aquaculture are discussed below:

1) **Pond culture:** Pond cultures are the most common mode of culture techniques. The number of ponds may range from one to hundreds, depending on the size of the harvest. The most common form of aquaculture pond is an earthen pond, which is made up of mainly of earth levees, while other materials can also be used. The pond size varies depending upon its purpose. A single pond can produce between 2,000 to 10,000 pounds per acre annually depending upon the production level.
2) **Cage culture:** Cage culture involves growing of aquatic organisms within cages, the size of which depends upon the need. The cage culture technique does not require the production of a pond-like structure. These can be placed in any water body which contains an inflow of water. One drawback of this technique is that these are susceptible to biological pollution, where non-native species being cultured escapes and disrupts the ecosystem.
3) **Raceway culture:** Raceway cultures are also called tank cultures. The tanks are made up of cement, fiberglass or any other durable substance. This is a more sophisticated as well as expensive procedure compared to the pond culture and cage culture. A constant water flow must be maintained, which regularly flushes out the wastes.
4) **Recirculating system:** Recirculating systems are the most expensive culture methods among all others. This process is almost similar to the raceway culture, except that the flow of water is controlled by pumps. The water is returned by the pumps to the reservoirs where the wastes are extracted and then the water is pumped back to the tanks for recycling.

3.2 PREPARATION OF COMPOUND DIET FOR FISH

The rapid worldwide expansion of aquaculture and livestock production strongly indicates that a crisis will be precipitated in the livestock and aquaculture feed industries in the near future. Fish are probably the most efficient converters of feed to the flesh, requiring from 2 to 4 lb of basic feedstuffs to produce 1 lb of fish. In contrast to land animals; however, fish are fastidious eaters in that they require higher levels of dietary protein. In addition, the amino acid requirements to promote rapid growth of fish appear to be more rigid than for land animals. Cultured fish require nutrients in their diet for their growth, reproduction, and other normal physiological functions. The common nutrients required for proper growth of fish are:

1) **Energy yielding nutrients:** Proteins, carbohydrates (10-30%) and lipids (4–28%).
2) **Protein and amino acids (32-45%):** Proteins compose approximately 70% of dry weight. Thus, protein is one of the most important nutritional compounds of fish feeds. A balanced mixture of amino acids like arginine, histidine, isoleucine, leucine, lysine, methionine, etc. is critical for adequate growth and health.
3) **Lipids:** Neutral lipids (fats and oils), in the form of triglycerides, fatty acids like linolenic acid, linoleic acid, etc. are also important for fish.
4) **Carbohydrates:** Soluble carbohydrates such as starch are essential for fish as it is the principal source of energy.
5) **Minerals (1.0-2.5%):** Fish requires around 20 inorganic mineral, including the macro minerals like calcium, phosphorus, magnesium, chloride, sodium, potassium, etc. and micro minerals like cobalt, chromium, copper, iodine, iron, manganese, zinc, etc. for their proper growth.
6) **Vitamins (1.0–2.5%):** Fifteen vitamins, including fat-solubie (A, D, E, and K) and water-soluble (C, B-complex consisting of thiamin, riboflavin, niacin, biotin, folic acid, choline, etc.) are essential for fish growth.

Often the natural food in a pond is insufficient for adequate growth and hence artificial feeding is necessary for successful pisciculture. Different types of artificial feed are:

1) **Vegetable feeds:** Agricultural wastes, including oil cakes and mill residues, plants, pulses, cereals, tubers, kitchen waste, etc.
2) **Feeds of animal origin**: Slaughterhouse refuse, fishery by-products including trash fish and fish meals, etc.
3) **Formulated feeds:** Formulated or compounded feeds may be given to fish in the dry, moist or wet form. The general composition of

compounded feed is proteins, fats, carbohydrates, cellulose, minerals, vitamins, binders and water. It is prepared by mixing (Royes and Chapman 2015):

- **Proteins and amino acid source:** Fish meal, soybean meal, fish hydrolysate, skim milk powder, legumes, wheat gluten, etc.
- **Lipids source:** Oil from vegetables like canola, sunflower, and linseed or from marine fish, like menhaden.
- **Carbohydrates:** Flours of corn, wheat or other cereals, rice bran, broken rice.
- **Vitamins and minerals:** Usually prepared synthetically.
- **Pigments:** Synthetic pigments like astaxanthin often used to enhance coloration in the salmon flesh or the skin of freshwater fish. Besides this, natural pigment source includes blue-green algae such as spirulina, dried shrimp meal, extracts from marigold, red peppers, etc.
- **Binding agents:** Beef heart, gelatin, gum arabic, locust bean, agar, carrageenans are used to provide stability to the pellet. Binding agents also reduce the leaching of nutrients into the water.
- **Preservatives:** Antimicrobials and antioxidants substances like butylated hydroxyanisole (BHA), butylated hydroxytoluene (BHT), ethoxyquin, etc. are generally added to extend the shelf-life of fish feed.
- **Attractants:** Attractants and flavourings, like fish hydrolysate, condensed fish soluble, glycine and alanine, betaine, etc. are often added to stimulate feeding behaviour in fish.

The preparation of formulated fish feeds begins with the formation of a dough-like mixture of above-mentioned ingredients. The dry ingredients obtained from feed stores are finely ground and mixed. A suitable volume of water is then added to produce the desirable consistency according to the type of fish. The last step is pelleting or rolling which converts the dough into pellets or flakes respectively.

During preparation, care must be taken on the amount of water added and heat because excess water in the mixture results in a soft pellet, while a temperature above 92^0C adversely affects the proteins and vitamins. Several formulated diets for fish larvae have been tested with relative success. However, the results on growth and survival achieved with these diets were inferior to the use of live food, possibly because of the composition, palatability, physical characteristics or to an inability of properly digesting the feed (Kolkovsky 2001).

The importance of using combined live and formulated diets lies mainly in supplying a more suitable and well-balanced diet to fish larvae in a digestible form. A number of other important aspects related to nutrition, species, fish biology, and husbandry conditions, which determine the success of using a combined feeding, are not yet fully understood. For instance, there appear to be specific periods during development when fish larvae will feed on, and metabolise formulated diets, and that can be accounted for a behavioural and a physiological capacity (Cahu and Zambonino-Infante 2001; Petkam and Moodie 2001). Moreover, some fish species larvae apparently are better adapted to the utilisation of nutrients from formulated diets than others.

Formulation of the compound, adequate diets for fish larvae at the early life stage is not easy to be achieved because the estimation of nutritional requirements of fish larvae cannot be carried out by traditional nutritional approaches. Moreover, live diets ingested by larvae contain exogenous substances such as gut neuropeptides, enzymes and nutritional growth factors that contribute to the digestion of prey, frequently omitted in formulated diets.

Zooplankton, a group of organisms ordinarily present in aquaculture ponds, is an essential diet for fish larvae, supplying many nutrients (in the digestive form) required by the larvae. A fish diet must provide a suitable energy source and be in proper balance with respect to all nutritional requirements like proteins, minerals, lipids, carbohydrates, vitamins and growth factors.

3.3 FISHING CRAFTS AND GEARS

The use of crafts and gears in fishing technology plays an important role in enhancing the production on a commercial basis. The success of fishing largely depends on to how and which types of nets are used to capture the fish. There are two main types of devices used to capture fish in both marine and inland fisheries: (1) Nets or gear- these are instruments used for catching fish and (2) Crafts or Boats- it provides a platform for fishing operations, carrying the crew and fishing gears.

There are various types of gears and crafts used in different parts depending upon the nature of water bodies, the age of fish and their species. Some nets are used without craft; however, others are used with the help of crafts. Generally, locally made gears and crafts may be non-mechanised or mechanised.

3.3.1 Fishing crafts

Fishing crafts are highly essential for catching fishes in large scale in water bodies. A variety of crafts (boats) are available and these can be divided broadly as non-mechanised and mechanised fishing crafts (Emmanuel 2010; Najmudeen and Sathiadhas 2008). Some of these are discussed below:

3.3.1.1 Non-mechanised boats

Some of the non-mechanised types of fishing crafts are:

- **Catamaran:** This is the simplest form of boat formed by joining a few curved logs and making it into a raft like structure that can float freely on the water surface.
- **Dug-out canoes:** These are simple fishing crafts made for fishing within short distances from the coast. These are formed by joining logs of wood in the form of a boat.
- **Plank-built canoes:** These are the enlarged versions of dug-out canoes which contain planks on the side of the canoes.
- **Masula boats:** These are made of non-rigid planks sewn together with coir ropes.
- **Dhinghi:** This is a carvel type of boat designed to perform various purposes including fishing.

3.3.1.2 Mechanised boats

Some of the mechanised types of fishing crafts are:

- **Hand line boat:** Hand lines boats operate all over the world. These can operate in both shallow and deep water. The traditional hand line boats do not use any winch.

- **Pole and line fishing vessels:** Pole and line fishing vessels are fitted with a narrow platform protruding all around the vessel at deck level, outside the bulwarks. The platform extends from the stern to the fore-end like a bowsprit.
- **Trolling vessel:** Trolling line boats tow extending on either side to catch pelagic species having high individual value and good quality.
- **Dol netter:** The dol netters are used to operate the doll nets, which are basically fish bag nets. The dol netters vary in length from 8-14mm, 1.5-3.6 mm in breadth and 0.8-1.8 mm in height.

3.3.2 Fishing gears

Fishing gears are the devices used in the aquatic bodies to capture different sizes of fish species. They may be of different size, shape and designs. These gears are made either by fishermen or manufactured in cottage industries. The various types of materials used to make fishing gears include netting, twine, plastic structural and fasteners, clips and swivels, ropes, steel wire ropes, combination wire ropes, purse rings, polyester, polyethylene, nylon, cotton, polypropylene, mixed fibers, floats and sinkers, bamboo, wood (Raju et al. 2016). Broadly the fishing gears are classified into two types - active gears and passive gears; which includes the following (Pauli and Sih 2017):

A] Active Gears: Fish capture by active gears is based on the aimed chase of the target fish species and then catching it with a suitable method. Active fishing methods are especially suitable for sampling large proportions of the whole fish stock or large numbers of fish. The term "active" means that the fishing gear is dragged through the water by human, animal or engine power.

1) **Seine:** Seines are fishing net, rectangular in shape mounted on a wire and hangs vertically in the water. They are generally set in a semicircle and dragged over a smooth bottom by means of long ropes. (Eyo and Akpati 1995**).** Seines are of following types:
    - **Beach seine:** Beach seines are especially used for catching seasonal pelagic species as they feed near the shore. The net is used in such a way that one fisher remains on the shore with one end of the net, while the other fisher on the beach holds the rest end of the net. The bottom and water surface act as natural

barriers which prevent the fish from escaping from the area enclosed by the net.

- **Boat Seines:** Boat seines are set and hauled from a boat. One end of the seine-is anchored and the fisher sails in a circle, releasing the net and returns to the anchor.
- **Purse Seines:** Purse seines are used to capture pelagic and migratory species. Purse seines are characterised by a floating line, which remains on the surface and a line at the bottom of the net that is used to close off the escape route. Purse seines are costly and required highly skilled operators.

2) **Trolling:** Trolling is a fishing method used to catch the pelagic fishes. Trolling lines are simple hooks that are trailed from a moving vessel at a controlled depth. (Eyo and Akpati 1995). Bait used for the target species may be made up of artificial or natural items. Trolling does not require high skills or a large investment in gear.

3) **Jigging:** The jigging technique involves mainly catching fish with special hooks. Generally, sharp hooks are used so that when they are pulled up it can penetrate the fish skin. Jigging is a low-cost, low–energy technique that does not require baits. Jigging requires knowledge of the local area to determine where and when it can be used.

4) **Cast nets:** Cast nets are conical-circular nets used to catch fish from the pelagic to the benthic section of the water body. When the net is thrown to the water surface the outer edge sinks to the bottom; thus, it is pulled and the fishes are collected. It is most useful in catching tilapia, which avoids most gears.

5) **Drift gill nets:** Drift nets are normally left to drift freely with the current. Drift nets are most often used when the flood is receding and during the dry season when the water level is low and the water current is not fast.

6) **Trawl nets:** Trawling is a method to catch a fish by pulling the fishing net through the water behind one or more boat. The net used for trawling is called trawl nets. Beam trawls are the simpler trawls and are used to capture flatfish and shrimp. Otter trawl nets are the major nets used for demersal fishing, while the Spanish trawl is a very large net and can be used in deep water down to 600 m on the continental slope.

7) **Clap nets:** Clap nets are commonly used in dry season to catch fish in pools or swamps. Clap nets are made up of wooden frames which contain 10 to 30 mm mesh size netting material. The fisher upon seeing the fish dives and claps the fish using the twin clap nets (one held in his right hand and the other in his left hand).

B] Passive Gears: Passive gears are in general the most ancient type of fishing gears. These gears are more suitable for small-scale fishing and are, therefore, often the gear types used in artisanal fisheries. Some passive fishing gears are often referred to as "stationary" fishing gears. Stationary gears are those anchored to the seabed and they constitute a large group of the passive gears. However, some moving gears such as drift nets may also be classified as passive gears, as fish capture by these gears also depends on the movement of the target species towards the gear.

1) **Gill net:** A gill net is made up of fiber netting. In the gill net, the fish is caught by the mesh of net, while trying to swim through. When the fish enters its head through the mesh and attempts to free itself, the twine slips under the gill cover, preventing escape. Trammel nets have three panels of netting The two outside walls of netting have a mesh size larger than the targeted fish and the interior netting has a smaller mesh size. The fish strikes large mesh panel, while passing from either side, forming a sac or pocket in which the fish are trapped.
2) **Hooks and lines:** This is the simplest gear employed for fishing. Hooks vary enormously in shape, size, type of point and the thickness of the wire. Various types of hooks are used, such as chain hooks, baited hooks for capturing larger fishes.
3) **Bag net:** It is a bag-like net made up of nylon with a mesh size of 2 - 3 mm and depth is of 0.5 - 1m provided with a circular iron frame and a long handle made with bamboo poles. Bag nets are kept vertically open by a frame (Raju et al. 2016).
4) **Trap nets:** It is used for fishing in shallow waters. The lower part is cylindrical, while the upper part is conical in shape. Interior region of the net contains cone-shaped necks to prevent the escape of fishes.
5) **Hand lift net:** It is a small, portable hand-operated net generally used in the shallow region of the river to catch small sized fish. It is rectangular in shape supported by a bamboo frame having two poles. Four corners of the net are tied into four corners of the bamboo frame. This gear is dipped in shallow waters for some time and then

lifted up rapidly to catch the entrapped fish which happen to be over the net by hand picking.

3.4 DEPLETION OF FISHERIES RESOURCES

Massive increase in fish demand has caused major fishing areas around the world to become over-exploited. There are no longer enough fishes to reproduce and rebuild the population. Most of the problems associated with overfishing have been caused in the last 50 years by the rapid advances in fishing technology.

Over consumption undoubtedly drives ecological impact, but it is by no means the only factor in the depletion of fish resources. The majority of global fish stocks lacks adequate catch, survey, and other biological data to calculate the current abundance and productivity using conventional stock assessment methods. In developed countries, the fraction of fish stocks that are assessed ranges between 10 and 50%. This fraction is generally lower in developing countries, where it ranges between 5 and 20% (Costello et al. 2012). This poses a significant challenge for the sustainable management of these stocks. Globally, fishing effort has expanded massively since the 1950s. A 2.4 times increase in yield has been achieved by a fourfold expansion in the fishing area (Swartz et al. 2010). This expansion extended from the coastal waters of the North Atlantic and West Pacific to the open-ocean, southern hemisphere and tropics. By the mid 1990s, one-third of the world's ocean and two-thirds of the continental shelves were exploited at levels, where more than 10% of marine primary production was appropriated to support fisheries (Swartz et al. 2010) and, in some cases of overfishing or overexploitation, these values can be exceeded for limited periods of time (Watson et al. 2013).

If everything continues as it has, all fisheries will continue or become over-exploited and eventually, the world's marine life can no longer survive. Populations are dropping so fast that the remaining fish cannot reproduce enough to sustain themselves. There are a number of reasons for the depletion of fisheries resources. Some of them are summerised below-

1) **Overfishing or overexploitation:** It is the condition of a water body (ponds, rivers, lakes or oceans) in which fish stocks are reduced to a level below acceptable limits. It occurs when fish are caught by fisherman to a limit which cannot be replaced or balanced through natural reproduction. There is always an incentive for all fishermen to catch as much as they can (Prisoner's dilemma). Overfishing results in resource depletion, reduced biological growth rates and low biomass levels. Globally, around 85% of the world's fisheries are over-exploited, depleted, fully exploited or in recovery from exploitation (Gaia 2002).

   According to a recent report (Living blue planet report), jointly published by World Wildlife Fund and the Zoological Society of London, there was a dramatic fall of 74% in worldwide stocks of the important scombridae fish such as mackerel, tuna and bonitos between 1970 and 2010 due to overexploitation (WWF, 2015).

2) **Organic load interference:** Eutrophication is linked to excessive amounts of organic loads (nitrogen and phosphorus rich compounds) from wastewater treatment plants, industrial facilities, agricultural runoff, stormwater, combined sewer overflows, urban runoff, and runoff from natural systems such as forests.

   Eutrophication is the depletion of oxygen level in a water body or fishery resource, which kills aquatic animals including fish. It occurs mainly due to excess nutrient enrichment, which in turn induces explosive growth of plants and algae, decaying of which consumes oxygen from the water.

3) **Erosion:** Deforestation and weathering of exposed soils results in soil erosion. This causes release of sediments or silt into the water body. Besides this, destructive fishing techniques like bottom trawling, dynamiting, and poisoning destroy habitats near shore as well as in the deep sea.

4) **Mesh size of fishing net:** Fishing gear system also contributes to the depletion of fishery resources. Thus, sustainable capture of fish from wild sources is very important. This can be achieved by limiting the size of nets, or by increasing the size of the mesh to avoid catching of juvenile fish, regulation of gear to avoid over exploitation of certain species; reservation of zones to traditional fishermen and declaration of closed seasons. For example, in Gujarat

- The area up to 9 km from the shore is reserved for non-mechanised vessels and mechanised vessels beyond 9 km.
- In case of trawl nets, square mesh of minimum 40 mm size at cod ends need to be used.
- Gill net with a mesh size less than 150 mm cannot be operated.

5) **Climate change:** In general, climate change causes warming of waters which in turn brings changes to rainfall patterns, water levels, river flow and water chemistry. Fish cannot control their body temperature, and thus increasing or decreasing water temperatures will have a significant adverse impact on growth and reproduction of fish (poor fecundity and even a failure to reproduce at all).

    For example, some fish like salmon, catfish and sturgeon cannot spawn if winter temperatures do not drop below a crucial level. On the other hand, higher temperatures will reduce oxygen levels, which make the water uninhabitable for many fishes.

6) **Weed infestation:** Infestation of floating aquatic weeds like water hyacinth also adversely affects the fishery resources. For example:
    - In Lake Victoria, a warm water temperature and disappearance of fish has been noted where there is much water hyacinth infestation. However, crocodiles and snakes have become more prevalent in this area.
    - Tilapiine and Nile tilapia prefer to breed and spawn in shallow in shore and sheltered areas. But, this spawning and nursery grounds are no longer available due to water hyacinth mats.
    - Mass fish mortality often occurs in the anoxic condition developed due to the decay of these weeds.
    - Excessive shading and low DO due to weed infestation also lower the primary productivity. Weeds also compete for nutrients with other plants on which some fish feed.

7) **Anthropogenic activity:** Direct or indirect dumping of excessive sewage, industrial effluents and toxic materials into the rivers and coastal sea adversely affects fishery resources. Due these anthropogenic activities, the country's major river systems are polluted, which results in large-scale fish kills and destruction of aquatic life. The high concentration of heavy metals like Zn, Cu, Mn, Hg, Pb, etc. released from industries into water body has significant negative impact on growth and reproduction of fish.

In addition to this, rival and powerful mechanised fishing vessels such as trawlers and purse seiners are responsible for destruction of fish breeding grounds on the sea bed, leading to severe depletion of marine stocks. Poisoning, electric fishing, padal, kolli are also very destructive types of fishing activities as they cause habitat destruction.

8) **Impact of agriculture:** Pesticides, weedicides used in agricultural fields are washed off to the downstream regions. Pesticides are also responsible for fishery resource depletion as they cause low disease resistance, reproductive success, sterility, reduced egg production, etc. in fish. Pesticides also reduce the availability of plants and insects that serve as habitat and natural fish food.

9) **Encroachment, sand mining, and construction of dam, bunds and barriers:** Artificial barriers like dam and bunds alter the prevailing ecological conditions of the natural habitat of fishes and thus break off the natural life cycle of fish species. Mangrove sanctuaries are important nursery grounds for many fish like mullet, sea bass, pearl spot or chromide fish, etc. Destruction of these sanctuaries has adversely affected fishery resources of this region (Kurup 2009).

3.5 APPLICATION OF REMOTE SENSING (RS) AND GIS IN FISHERIES

Planning for aquaculture development requires an understanding of the environment and assessing the suitability of a given region or site for a project to be sustainable. Water quality and physical properties of the water bodies can be assessed by remote sensing (RS), with some limitations related to the complexity of coastal environments. Applications of RS and Geographical Information System (GIS) technologies have increased dramatically since the mid-1980s (Meaden 2013). Although GIS and remote sensing have been widely applied to marine fisheries, there have been fewer applications of these technologies in inland fisheries management and planning. Like many marine fisheries GIS applications, inland fisheries applications of GIS have largely dealt with mapping the distribution and abundance of fish species, and mapping and modelling habitat in rivers, reservoirs and lakes, and relating the two.

3.5.1 Remote Sensing (RS)

Remote Sensing (RS) refers to the branch of science which derives information about objects from measurements made from a distance i.e. without actually coming into contact with them. Conventionally remote sensing deals with the use of light, i.e. electromagnetic radiation as the medium of interaction. RS refers to the identification of earth features by detecting the characteristics electromagnetic radiation that is reflected from the earth's surface. Every object reflects a portion of the electromagnetic radiation incident on it depending upon its physical properties. In addition, objects also emit electromagnetic radiation depending upon their temperature and emissivity. Reflectance pattern at different wavelengths for each object is different. Such a set of characteristics is known as the spectral signature of the object. This enables identification and discrimination of objects. Visual perception of objects is the best example of remote sensing. The common applications of remote sensing in fisheries are summarised below:

1) Remote sensing data help in the regular management of water resources.
2) Remote sensing techniques are useful in finding different types of bioresources. It plays a potential role in both rapid and comprehensive EIA.
3) For detection and monitoring of the water pollution, remote sensing proves useful.
4) Remote sensing is applicable in acquiring information regarding offshore engineering activities, fisheries surveillance, ocean features, coastal regions and storm forecast operations.
5) Remotely sensed data provides the necessary spatial data on suspended sediments, dissolved organic matter, phytoplankton, algal blooms and oil slicks, etc., which will useful in the management of fish stocks, monitor the water quality and natural water pollution such as oil spill or algal blooms, which are harmful to aquatic life.
6) Remote sensing techniques are giving necessary data needed for monitoring changes on coastal erosion, shoreline monitoring and

management, loss of natural habitat, sea level rise, wetland mapping urbanisation, sewage disposal and aquatic population, etc.

7) Remote sensing is very useful in identifying Potential Fishing Zones (PFZ). This data is very useful for fishermen because they came to know likely occurrence of fish shoals which helps them for getting more catch.
8) Continuous monitoring of land use or land cover with remote sensing imageries has been of immense use in providing information on temporal and spatial changes in the area under aquaculture, mangrove areas, coral reef mapping and other land use patterns.

3.5.2 Geographical Information System (GIS)

GIS may be defined as the integration of computer hard and software with spatially referred digital data so that storage, retrieval, manipulation, analysis and display all forms of geographically referenced information. In other words, GIS is a computer assisted system that can input, store, retrieve, analyse and display geographically referenced information useful for decision making. The definition of GIS is not that important, but it must encompass

- Data and concepts concerned with spatial distribution (Geographical).
- Notion of conveying data, ideas or analysis (Information).
- The sequence of inputs, processes and outputs (System).

GPS is becoming popular with its diversified fields of applications including fishery. The common applications of GPS in fisheries are summarised below-

1) Identification of suitable sites for freshwater and brackish water aquaculture.
2) Management of marine fisheries and coastal regulation zone.
3) Study of the land-use pattern, including mangroves and forest cover of a particular area.
4) Planning for water body resource zonation and mapping of aquatic species.
5) Fish disease modelling and management.
6) Study of temporal/spatial changes in fish production and consumption.

7) Environmental impact assessment.
8) The distribution of different fish species in relation to physical habitat characteristics and study of spatial variations in demand/supply balance.

3.6 LAWS AND REGULATIONS IN FISHERIES

Fisheries law refers to state and federal legislation regarding the protection of endangered species of fish and the protection of their habitats, as well as statutes designed to ensure the safety of fish products used by consumers. Fisheries law also refers to federal and state laws regulating commercial and sport fishing activities, such as fishing licenses, permits, catch limits and the dates of the fishing season (https://www.hg.org/).

Fisheries experts in Southeast Asia now recognise that a fishery cannot be managed effectively without the cooperation of fisherman to make laws and regulations work. The constitution of India favours a political structure in which the power of enacting laws is split between India's central government and the Indian states. Several key laws and regulations (central level) which are relevant to fisheries and aquaculture include-

1) **Indian Fisheries Act (1897):** The present Act extends to the whole of India, as it was before independence. It is divided in 7 sections:
   a) **Title and extent:** (1) This Act may be called the Indian Fisheries Act, 1897. (2) It extends to the whole of India except the Part B States.
   b) **Act to be read as a supplement to other fisheries laws:** Subject to the provisions of sections 8 and 10 of the General Clauses Act, 1887, this Act shall be read as supplemental to any other enactment.

   For law relating to Fisheries in-

   (1) Assam, see the Assam Land and Revenue Regulation, 1886 (1 of 1886), ss.16 and 155;

   (2) Bengal and Assam (Private Fisheries), see the Private Fisheries Protection Act, 1889 (Ben.2 of 1889);

   (3) Nilgiris District, as to acclimatised fish, see the Nilgiris Game and Fish Preservation Act, 1879 (Mad.2 of 1879);

(4) Punjab, see the Punjab Fisheries Act, 1914.

c) **Definitions:** In this Act, unless there is anything repugnant in the subject or context-

(1) "fish" includes shell-fish:

(2) "fixed engine" means any net, cage, trap or other contrivance for taking fish, fixed in the soil or made stationary in any other way: and

(3) "private water" means water which is the exclusive property of any person or in which any person has for the time being an exclusive right of fishery whether as owner, lessee or in any other capacity.

Explanation.-Water shall not cease to be "private water" within the meaning of this definition by reason only that other persons may have by custom a right of fishery therein

d) **Destruction of fish by explosives in inland waters and on coast:**

(1) If any person uses any dynamite or other explosive substance in any water with intent thereby to catch or destroy any of the fish that may be therein, he shall be punishable with imprisonment for a term which may extend to two months, or with fine which may extend to two hundred rupees.

(2) In sub-section (1) the word "water" includes the sea within a distance of one marine league of the sea-coast: and an offence committed under that sub-section in such sea may be tried, punished and in all respects dealt with as if it had been committed on the land abutting on such coast.

e) **Destruction of fish by poisoning of waters:**

(1) If any person puts any poison, lime or noxious material into any water with intent thereby to catch or destroy any fish he shall be punishable with imprisonment for a term which may extend to two months, or with fine which may be extended to two hundred rupees.

(2) The State Government may, by notification in the Official Gazette, suspend the operation of this section in any specified area, and may in like manner modify or cancel any such notification.

f) **Protection of fish in selected waters by rules of state government:**

(1) The State Government may make rules for the purposes hereinafter in this section mentioned, and may, by notification in the Official Gazette, apply all or any of such rules to such waters, not being private waters, as the State Government may specify in the said notification.

(2) The State Government may also, by like notification, apply such rules or any of them to any private water with the consent in writing of the owner thereof and of all persons having for the time being any exclusive right of fishery therein

(3) Such rules may prohibit or regulate all or any of the following matters, that is to say-

(a) The erection and use of fixed engines;

(b) The construction of weirs; and

(c) The dimension and kind of the nets to be used and the modes of using them.

(4) Such rules may also prohibit all fishing in any specified water for a period not exceeding two years.

(5) In making any rule under this section, the State Government may

(a) Direct that a breach of it shall be punishable with fine which may extend to one hundred rupees, and, when the breach is a continuing breach, with a further fine which may extend to ten rupees for every day after the date of the first conviction during which the breach is proved to have been persisted in; and

(b) Provide for- (i) The seizure, forfeiture and removal of fixed engines, erected or used, or nets used, in contravention of the rule, and (ii) The forfeiture of any fish taken by means of any such fixed engine or net.

(6) The power to make rules under this section is subject to the condition that they shall be made after previous publication.

g) Arrest without warrant for offences under this Act:

(1) Any police-officer, or other person specially empowered by the State Government in this behalf, either by name or as holding any office, for the time being, may, without an order from a Magistrate and without warrant, arrest any person committing in his view any Act offence punishable under section 4 or 5 or under any rule under section 6-

(a) If the name and address of the person are unknown to him, and

(b) If the person declines to give his name and address; or if there is a reason to doubt the accuracy of the name and address if given.

(2) A person arrested under this section may be detained until his name and address have been correctly ascertained:

2) **Environment Protection Act (1986):** It is an umbrella act containing provisions for all environment related issues affecting fisheries and aquaculture industry in India. The purpose of this act is the protection and improvement of the human environment and the prevention of hazards to human beings, other living creatures including fish, plants and property.
3) **Water (Prevention and Control of Pollution) Act, 1974:** It applies to the whole of the states of Assam, Bihar, Gujarat, Haryana, Himachal Pradesh, Jammu and Kashmir, Karnataka, Kerala, Madhya Pradesh, Rajasthan, Tripura and West Bengal and the Union territories. It shall also apply to such other states of India, which adopts this Act by resolution passed in that behalf under clause (1) of article 252 of the constitution. The aim of water Act is the prevention and control of water pollution in India.
4) **Wildlife Protection Act, 1972:** It is an Act of the Indian Parliament acted out for the protection of plants and animal species. It provides protection of wild animals (including fish) with a view to ensuring ecological and environmental security. It extends to the whole of India, except the state of Jammu and Kashmir, which has its own wildlife act. According to this act, "wildlife" includes any animal, bees, butterflies, crustacean, fish and moths; and aquatic or land vegetation which forms part of any natural habitat.

Fisheries in the maritime states of India, within the territorial limits of 12 miles, are dealt with under the Marine Fishing Regulation Act (MFRA). This act is formulated on the guidelines provided by the model piece of legislation prepared by the Ministry of Agriculture, GOI, in 1979, which was encouraged by the fishers operating unpowered fishing vessels to safeguard their fishing space and equipment from bottom trawlers. Currently, these legislations are not just restricted to the maritime states of the country, but are quite widespread in other states as well, for regulating fisheries in inland waters. This development, in particular, can be seen as a positive initiative by other states that has and will help in taking such welfare measures to a higher level (http://fishlaw.org/).

Some of the important management actions adopted under the MFRA are the prohibition on certain fishing gear, regulation on the mesh size of net, the establishment of closed season and areas, demarcation of zones for no trawling, etc. along with other measures such as the use of turtle excluder devices and designation of no-fishing areas. Other than this, there are specific provisions regarding appointment of adjudicating officers and other officials as per the need and requirement of the acts, issuing of licenses upon meeting several standards, registration of fishing vessels, regulating the manner of fishing with the least threat to the environment, penalties and fines in cases of non-compliance, and maintaining a balance between operators of mechanised boats and traditional fishermen using non-mechanised boats.

The aims behind such provisions were to conserve fish and to regulate fishing on a scientific basis, and maintain law and order in the sea and inland waters (http://fishlaw.org/). In addition, the legislations made by different states are quite successful in providing a proper framework for regulating this sector. However, there are still many questions being raised about the implementation of these laws at the field level, and this will remain a grey area for quite some time.

On 11 December 1996, the Indian Supreme Court had taken a historic decision in a case regarding the setting up of shrimp farms in coastal areas.

a) The Supreme Court prohibited the construction/set up of shrimp culture ponds within the Coastal Regulation Zone (CRZ) and within 1000 meters of Chilka Lake and Pulicat Lake, except traditional and improved traditional types of ponds.

b) Authority should be constituted to protect the ecologically fragile coastal areas, seashore, waterfront and other coastal areas, particularly to deal with the situation created by the shrimp culture industry in the coastal states/union territories.

To perform the functions indicated by the Supreme Court, the Ministry of Agriculture (aquaculture authority) issued guidelines for sustainable development and management of brackish water aquaculture. The primary responsibilities of aquaculture authorities include the following-

1) They recommend states to identify lands that are fit for aquaculture and simultaneously to discourage the conversion of

healthy agriculture land into water body for aquaculture (http://www.fao.org/fishery/legalframework/nalo_india/en).

2) The guidelines also recognise the importance of wastewater treatment and prescribe standards for the treatment of wastewater discharged from aquaculture systems, hatcheries, feed mills and processing plants.
3) It also formulated guidelines for adopting improved technology for increasing production and productivity in traditional and improved traditional systems with the objective of optimising yield levels on a sustainable basis.
4) The aquaculture authority has constituted state level committees (SLCs) and district level committees (DLCs). Farmers who are operating traditional and improved traditional systems of aquaculture within the coastal regulation zone or within 1000 metres of Chilka Lake and Pulicat Lake for increased production and productivity are required to apply for prior approval for adoption of improved technology. The aquaculture authority recently drafted guidelines for effluent treatment system in shrimp farms.

Besides this, laws and regulations are there many laws and regulation formulated by both state and central government of India, which are either directly or indirectly related to aquaculture or fishery practices. For example:

1) The Environmental Impact Assessment Notification (1994), in accordance with the Environment (Protection) Act, specifies the industries and projects (listed in Schedule I) that require an EIA. It also states that applications must be submitted to the Ministry of Environment and Forests and assessments should be completed within a period of 90 days.
2) The Prevention of Cruelty to Animals Act (1960) prevents the infliction of unnecessary pain or suffering on animals including fish. The breeding of and experiments on animals rules of this Act state that no establishment shall a) carry on the business of animal breeding, b) perform experiments like transgenic or genetic manipulation on animals and c) trade animals for the purpose of the experiments, unless it is registered.
3) The Export (Quality Control and Inspection) Act (1963) states that fresh, frozen and processed fish and fishery products must be

subject to quality control, inspection and monitoring prior to export. Maximum residual limits (MRLs) for pesticides, heavy metals and antibiotics or other pharmacologically active substances in fish and fishery products must be checked prior to export (FAO, http://www.fao.org/fishery/legalframework).

4) The Export of Fresh, Frozen and Processed Fish and Fishery Products (Quality Control and Inspection and Monitoring) Rules (1995) states that it is the primary responsibility of the industry to ensure that fish and fishery products or any other commodities intended for export are handled, processed at all stages of production, stored and transported under proper hygienic conditions so as to meet the health requirements.
5) The coastal regulation zone notification no. SO 114 (E) (1991), issued under the Environment Protection Act states that the entire coastal stretch of seas, bays, estuaries, rivers and backwaters from the lowest low tide to the highest high tide line and the coastal land within 500 m from the high tide line on the landward side are coastal regulation zone (CRZ). Within the CRZ the setting up or expansion of industries is prohibited.

3.7 REFERENCES

Cahu C and Infante JZ (2001) Substitution of live food by formulated diets in marine fish larvae. Aquaculture. 200(1): 161-180.

Carruthers T R, Punt AE, Walters CJ, MacCall A, McAllister MK, Dick EJ, Cope J (2014) Evaluating methods for setting catch limits in data-limited fisheries. Fisheries Research. 153: 48-68.

Costello C, Ovando D, Hilborn R, Gains SD, Deschenes O, Lester SE (2012) Status and solutions for the world's unassessed fisheries. Science. 338(6106): 517–520.

Das BK, Singh NR, Boruah P, Kar D (2015) Fishing devices of the river Siang in Arunachal Pradesh, India. Journal of Fisheries. 3(2): 251-258.

Emmanuel BE (2010) Fishing crafts characteristics and preservation techniques in Lekki lagoon, Nigeria. The Journal of American Science. 6(2): 18-23.

Eyo JE and Akpati CL (1995) Fishing gears and fishing methods. Proceeding of the UNDP assisted agriculture and rural development programme (Ministry of Agriculture, Awka, Anambra State). Training Course on Artisanal Fisheries Development, held at the University of Nigeria, Nsukka, 29th, Oct–12th, Nov, 147-167.

FAO, National Aquaculture legislation overview-India, http://www.fao.org/fishery/legalframework/nalo_india/en

Gabriel O, Lange K, Dahm E, Wendt T (2005). Fish catching methods of the world, 4th edn. John Wiley & Sons, Inc, Hoboken, New Jersey, pp. 1-536.

Gaia V (2002) BBC - Future - How the world's oceans could be running out of fish (http://www.bbc.com/).

Kolkovski S (2001) Digestive enzymes in fish larvae and juveniles—implications and applications to formulated diets. Aquaculture. 200(1): 181-201.

Kurup BM (2009) Coastal and marine fisheries resources of Kerala-conservation and sustainable development. In: Jayappa KS and Narayana AC eds. Coastal environments: Problems and perspectives. I.K. International Publishing House Pvt. Ltd., New Delhi, pp. 34-37.

Meaden GJ and Aguilar-Manjarrez J (2013) Advances in geographic information systems and remote sensing for fisheries and aquaculture. FAO Fisheries and Aquaculture Technical Paper No. 552, FAO, Rome, Italy, 98 p.

Mujere N (2015) Water Hyacinth: characteristics, problems, control options, and beneficial uses. In: McKeown AE and Bugyi G eds. Impact of Water Pollution on Human Health and Environmental Sustainability. Information science reference, IGI Global, Hershey, USA, pp. 346-347.

Najmudeen TM and Sathiadhas R (2008) Economic impact of juvenile fishing in a tropical multi-gear multi-species fishery. Fisheries Research. 92(2): 322-332.

Olaniyan RF (2015) Fishing Methods and their Implications for a Sustainable Environment. Fisheries and Aquaculture Journal. 6:139.

Pauli BD and Sih A (2017) Behavioural responses to human-induced change: Why fishing should not be ignored. Evolutionary Applications. 10(3): 231–240.

Petkam R and Moodie GEE (2001) Food particle size, feeding frequency, and the use of prepared food to culture larval walking catfish (*Clarias macrocephalus*). Aquaculture. 194(3): 349-362.

Quinitio ET and Parado-Estepa FD (2003) Biology and hatchery of mud crabs *Scylla spp.* Aquaculture Department, Southeast Asian Fisheries Development Center.

Raju CS, Rao JCS, Rao KG, Simhachalam G (2016) Fishing methods, use of indigenous knowledge and traditional practices in fisheries management of Lake Kolleru. Journal of Entomology and Zoology Studies. 4(5): 37- 44.

Royes JAB and Chapman FA (2015) Fisheries and Aquatic Sciences Department, UF/IFAS Extension. http://edis.ifas.ufl.edu/fa097.

Stottrup J and McEvoy L (2003). Live feeds in marine aquaculture, 1st edn. John Wiley & Sons, Inc, Hoboken, New Jersey, pp. 1-12.

Swartz W, Sala E, Tracey S, Watson R, Pauly D, Sandin SA (2010) The spatial expansion and ecological footprint of fisheries (1950 to present). PLoS One. 5: e15143.

Tucker CS and Hargreaves JA (2004) Biology and culture of channel catfish, 1st edn. Elsevier, Amsterdam, Netherlands, pp. 1-686.

Watson R, Zeller D, Pauly D (2013) Primary production demands of global fishing fleets. Fish and Fisheries. 15(2): 231–241.

Welcomme RL, Cowx IG, Coates D (2010) Inland capture fisheries. Philosophical Transactions of the Royal Society B: Biological Sciences. 365(1554): 2881-2896.

*** ***** ***

Chapter 4

AQUACULTURE PRACTICES

Joyobrato Nath

Department of Zoology, Cachar College, Silchar, Assam, India

Chapter highlights:

Aquaculture is an industrial process of raising of aquatic organisms such as fish, crustaceans, and molluscs including aquatic plants (seaweed) in all types of water environments under controlled conditions. It offered a wide scope ranging from production of protein-rich food in the era of food insecurity to socioeconomic development. Human population is growing day by day and simultaneously aquatic habitats are also destroyed bit by bit. The production of enough protein rich nutritious food sustainably to feed everyone is one of the major challenges of the present day. The concept of sustainable aquaculture (RAS, cage culture, pen culture, etc.) has evolved and grown in parallel with the growing evidence of overexploitation of wild fisheries and simultaneous extinction of many fish species. In this direction, improved water quality management practices, sewage fed fishery, composite fish culture and integrated fish farming will manage the present crises of getting animal protein from the existing water resources up to some extent. The fish aquarium is a source of employment and income for individuals/companies of different countries involved in its design, construction and distribution. Additionally, aquarium improves the scenic beauty of the environment and thus viewing certainly reduces stress and subsequently lowers blood pressure as well aid in controlling Alzheimer's disease.

4.1 AQUACULTURE AND ITS IMPORTANCE

4.1.1 Aquaculture or Aqua farming

Aquaculture is an industrial process of raising (breeding, rearing, and harvesting) of aquatic organisms such as fish, crustaceans, and molluscs including aquatic plants (seaweed) in all types of water environments under controlled conditions. In contrast, commercial fishing involves harvesting of wild fish from their native environment.

According to Food and Agriculture Organization (FAO) aquaculture is "understood to mean the farming of aquatic organisms, including fish, molluscs, crustaceans and aquatic plants. Farming implies some form of intervention in the rearing process to enhance production, such as regular stocking, feeding, protection from predators, etc."

4.1.2 Importance of aquaculture

Aquaculture offered a wide scope ranging from production of protein rich food in the era of food insecurity to socioeconomic development.

1) **Protein-rich food for human:** As the world's population is growing exponentially, the production of enough nutritious food sustainably to feed everyone is one of the major challenges of the present day. Production of protein through aquaculture is much more efficient compared to its production in any other animal production system.

   Compared to other animals, fish can consume more protein and can subsequently convert nitrogen in the feed into structural proteins making the fish more efficient producer of protein than the cow or the chicken. When a fish fed on a balanced diet under favourable environmental conditions a feed conversion rate (FCR) of 1:1 (kg body mass gain per kg feed intake) can be obtained in fish. For example, in channel catfish and rainbow trout, an FCR of 1.0 – 1.25 has been obtained. Similarly, the protein efficiency ratio (weight gain per unit of protein intake) is also higher for fish than for pig, sheep or steers. The reason behind low PFR in pig, sheep, etc. is the loss of amino acids through deamination (FAO 1987).

   In addition to the above, the higher efficiency of nitrogen excretion in fish (ammonotelic) is an added advantage from the bioenergetic point of view. They save energy as ammonia need not be converted into larger compounds using energy dependent urea cycle.

2) **Aesthetic appeal:** Ornamental fish keeping is one of the most popular hobbies in the world today. In this regard, ornamental fishes form an important commercial component of aquaculture for aesthetic requirements. The growing interest in ornamental fishes has resulted in a steady increase in global aquarium fish trade. An estimated trade with a turnover of about US $ 5 Billion and an annual growth rate of 8 percent offers a lot of scope for its sustainable development. In the global ornamental fish trade, India's share is about Rs 158.23 lakhs which is only 0.008% of the global trade (Shivta et al. 2013).

3) **Recreational fishing:** The traditional sports fishing is enjoyed by millions of people worldwide. Fishing primarily for pleasure or competition is called recreational fishing or sports fishing. Common fishing gears used in recreational fishing are rod, reel, line, hooks and any one of a wide range of baits. This leads to the development of recreational fishing industry associated with manufacture of fishing tackle, the design and building of recreational fishing boats.

4) **Socioeconomic value:** Both commercial and industrial aquaculture provides a means of earning the livelihood and monetary profit. Thus, fish resources have a major contribution to economic growth and poverty alleviation. For example, fishing in India contributed over 1 percent of India's annual gross domestic product in 2008, employing about 14.5 million people.

5) **Seaweed farming and economic development:** Seaweed (*Gelidium, Porphyra*, and *Laminaria*) farming has frequently been developed as an alternative to reduce fishing pressure. Seaweeds are also harvested or cultivated for the extraction of alginate, agar and carrageenan. Beside pisciculture, seaweed farming is an alternative to improve economic conditions. For instance, the annual production value of nori (*Pyropia*) is about US$2 billion in Japan and is one of the world's most valuable crops produced by aquaculture.

6) **Oyster culture and economic development:** Pearls are very valuable due to the high demand and prices for them. Today, more than 99% of all pearls sold worldwide are cultured pearls. In addition; edible oyster is one of the most widely cultivated bivalves. Thus, the contribution of oyster culture in a country's economic growth and also meeting the growing food demand cannot be neglected. Many countries, including India, import a lot of cultured pearls from International markets to meet the domestic demand.

4.2 SUSTAINABLE AQUACULTURE

The concept of sustainable aquaculture has evolved and grown in parallel with the growing evidence of overexploitation of wild fisheries and simultaneous extinction of many fish species. Briefly, sustainable aquaculture is the production of aquatic organisms using efficient and cost-effective methods to improve human capacity, resource use along with simultaneous conservation and protection of the available resources and the environment. Some essential practices employed in sustainable aquaculture include

1) **Environment practices:** Conservation of mangrove and wetland; effective effluent management and water quality control; efficient fishmeal and fish oil use; responsible sourcing of broodstock and juvenile fish.
2) **Community practices:** Establishment of well-defined rights and responsibilities for aquaculturists; community involvement; worker safety, fair labour practices and equitable compensation.
3) **Sustainable business and farm management practices:** Effective biosecurity and disease control systems; minimal antibiotic use; efficient harvest and transport; accountable recordkeeping; profitability.

On the basis of the level of inputs of feed and/or fertilisers and stocking density, fish farming is commonly categorised into the following three types:

1) **Extensive culture:** This type of farming receives no intentional nutritional inputs (manure and feed) and the fish completely depends on available natural food (zooplankton and phytoplankton) in the culture system. Thus, in this system, fish are grown without the use of fertiliser or farmer feeding. Extensive fish farming is generally done in the ocean, natural and man-made lakes, rivers and fiords.

   The main drawbacks of this system are a) it depends on the surrounding water for an overall mortality rate, survivorship and growth rate of the fish and b) high risk of algal blooms. The decaying algal biomass will deplete the oxygen in the extensive culture system results in massive loss of fish.

   To utilise all available food sources in the pond, the aquaculturist in an extensive system often choose fish species which occupy different niches like a filter algae feeder such as tilapia, a benthic feeder such as carp or catfish and a zooplankton feeder or submerged weeds feeder such as grass carp.

Despite these limitations, common carp and a number of other fish species are still farmed extensively in the European Union. In the Czech Republic, trout and carp are harvested each year from of natural and semi-natural ponds. In Asia and Africa, extensive culture of carp and tilapia is very common. For example, a common method of culturing bighead carp in China is to grow them in small lakes and reservoirs without the use of feed or fertiliser (Weimin 2004a).

2) **Semi-intensive culture:** Semi-intensive fish farming depends largely on natural food and manure and/or very little supplementary feed. They are given supplementary feed at least two times per week and the fertilisation is done once per week. The feed may be of vegetable origin or may include fish and fish oil. Thus, in semi-intensive fish farming fish can grow faster at a comparatively greater stocking density.

   In semi-intensive culture system, the grass carp are stocked in ponds or pens generally with other carp species, where aquatic weeds and pond side grasses form the main feed. Commercial feeds, such as pellets, and vegetable by-products are used when the growth of water grasses and algae is poor (Weimin 2004b).

3) **Intensive culture:** In intensive farming, fish are raised at very high stocking densities under controlled conditions and are subject to supplemental feeding and fertilisation. In this system, fish are dependent on the feed provided by the farmer such as fresh, wild, marine or freshwater fish, or on formulated diets, usually in dry pelleted form (Table 4.1).

   Along with feeding, this culture system also provides adequate water quality by maintaining optimal temperature levels, the level of pH, oxygen levels, stocking densities, the concentration of ammonia and periodically removing waste to promote growth, reduce stress, control disease, and reduce mortality. Fish farming in an intensive system become the easiest way of increasing the fish production. For example, in the EU, 80% of farmed fish production comprises of rainbow trout, marine fishes Atlantic salmon, gilthead seabeam and European seabass are predominantly intensively farmed.

   The cost of inputs per unit of fish weight is higher than in extensive farming, due to the high cost of fish feed. However, higher food conversion efficiency of aquatic animals including the fish makes it a good source of protein. E.g., salmon have an FCR of around 1.1 kg of feed per kg of salmon, while chickens have 2.5 kg of feed per kg of chicken (Torrissen et al. 2011). The intensive culture system aims to produce high-value fish at high stocking density. The three common approaches used in intensive culture system are:

a) **Land-based intensive flow-through farming:** In this method, water flows through the culture system only once and is then discharged back to the aquatic environment. The flowing incoming water supplies oxygen to the fish in the system and simultaneously carries wastes out of the system along with the outgoing water. This process is repeated at least once per day. Example- In Europe, most of the trout farming is done through this method.

b) **Recirculation Aquaculture systems (RAS):** RAS is also a land based system in which water is re-used after mechanical and biological treatment to reduce the load of dissolved and suspended wastes including ammonia. The main freshwater species produced in RAS are eel, trout and catfish. To maintain adequate water quality, a series of treatment processes are employed such as biofiltration, solids removal, oxygenation, pH control etc.

   This system presents several advantages such as reduced water requirements and land needs, high biosecurity levels, proper controlled environmental condition, easy removal of wastes including ammonia. A special type of RAS called **aquaponics** combines aquaculture (fish culture) and **hydroponics** (growing plants in water). In this type of system ammonia produced by the fish is removed by the plants (Chinese cabbage, lettuce, spinach, chives, herbs, etc.) from the water (Fig. 4.1). In this integrated system,

   - Water from an aquaculture system is carried by pump to a hydroponic system (Fig. 4.2).
   - In hydroponic system, by-products of aquaculture are broken down by nitrifying bacteria initially into nitrites and then into nitrates, which are utilised by the plants as nutrients. The ammonia or waste free water is then recycled back to the aquaculture system.

c) **Cage farming:** Rearing fish in cages is used in many parts of the world. In this method fish are stocked in cages, artificially fed, and harvested when they reach market size. It is also called **off-shore cultivation**, particularly when the cages are placed in the sea. The sea cages are widely used for rearing salmon, sea bass, sea bream, and trout.

This system presents several advantages such as unlimited water supply provides vast amounts of oxygen and running water, an easy and low cost of harvesting; fish feeding response and health can be monitored closely, etc.

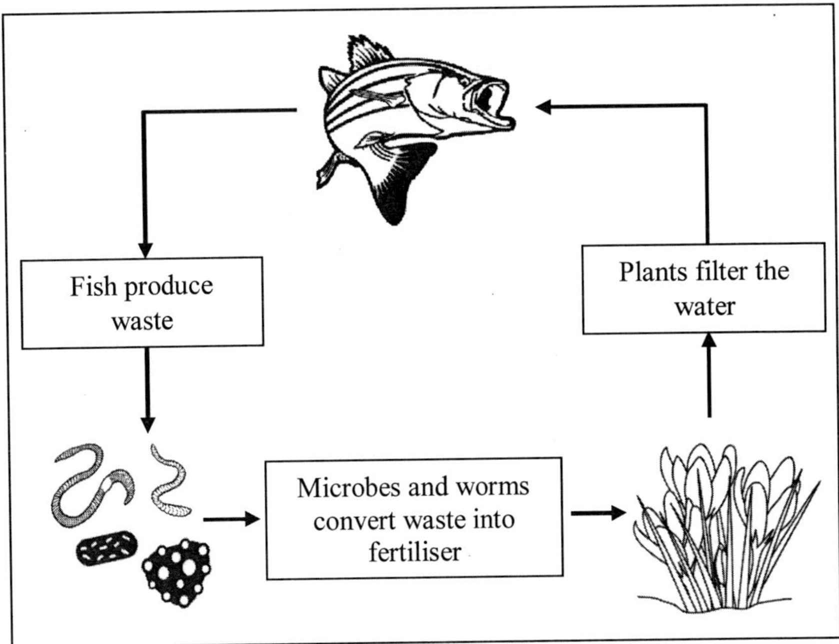

Fig. 4.1: The aquaponics cycle (aquaculture+ hydroponic) for fish farming.

Measures taken in intensive culture system: Some of the common measures taken in intensive fish farming are:

1) Aeration of the water is essential for productive fish farming. This is achieved by bubbling, cascade flow or aqueous oxygen.
2) Proper purification of water using approaches like biofiltration (conversion of toxic ammonia into less toxic nitrate by nitrifying bacteria), aquaponics, solids removal by a sand filter or particle filter.
3) Proper control of temperature as each fish has its own optimum temperature for maximum growth rate such as tilapia and

barramundi prefer around 24^0C water temperature, while trout and salmon show maximum growth at a water temperature below 16°C. Water temperature can be controlled by using submerged heaters, heat pumps, chillers, and heat exchangers (Odd-Ivar Lekang 2013).

4) The risk of infections by parasites like fish lice, fungi, intestinal worms, bacteria and protozoa is often higher in high population densities. This can be minimised by UV or ozone water treatment system. Recently, antimicrobial copper alloys are used in cage culture to prevent biofouling.
5) In intensive culture system, water pH must be monitored and controlled as it has a significant effect on growth rate. Water pH can be controlled by using lime or by degassing CO_2 in a packed column.

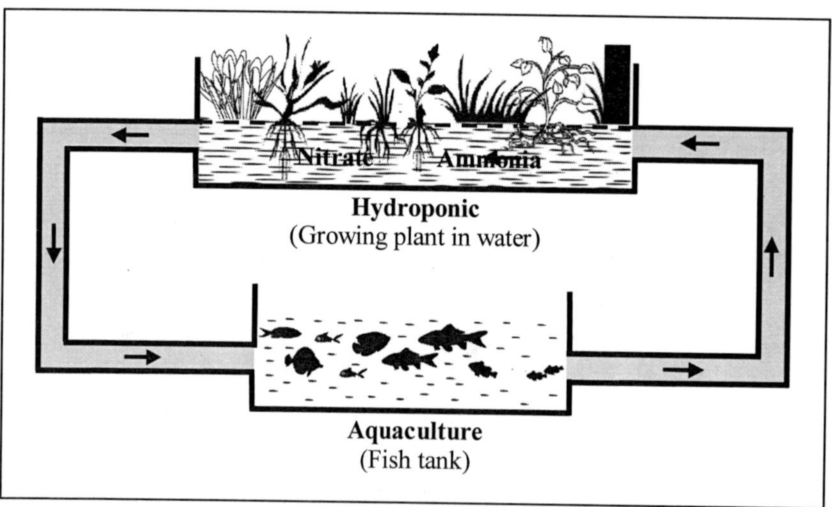

Fig. 4.2: A model of aquaponics for fish culture.

Table 4.1: Differences between extensive, semi-intensive and intensive culture.

Salient features	Extensive	Semi-intensive	Intensive
Nutritional input	No nutritional inputs (manure and supplementary feed)	Manure based culture practice	Formulated feeding (protein rich food)

Food	Fish depend on natural food available in the pond	Limited supply of supplementary food	Controlled environment (aeration, waste removal, etc.)
Stocking density	low	Moderate	High
Fish production	low	Higher than extensive	Highest per unit area compared to the other former two.
Production cost	low	Moderate	High

4.3 PEN CULTURE AND CAGE CULTURE

The terms cage and pen are often used interchangeably. For instance, in North America, 'sea pens' and 'sea cages' refer to the same culture method (Novotny 1975). Both pen culture and cage culture are a type of enclosure fish farming practice where fishes are cultured within an enclosed space, while maintaining a free exchange of water. The only difference is in their construction.

4.3.1 Cage culture system

A cage is fully enclosed on all sides or sometimes opens at the top, either floating on the surface or totally submerged in water (Fig. 4.3). In this method fish are stocked in cages, artificially fed, and harvested when they reach marketable size.

Fish cages can be placed in existing water resources like lakes, bayous, ponds, rivers or open oceans. It is also called off-shore cultivation, particularly when the cages are placed in the sea. The sea cages are widely used for rearing salmon, sea bass, sea bream, and trout. Cages are generally made of a rigid frame covered with a mesh from all sides. The mesh holds the fish within the cage, allowing the farmer easily to feed, observe and harvest them. The mesh allows the water to pass freely through the cage; thus, overcome the depletion of dissolved oxygen and minimises the chance of waste (ammonia) accumulation. Many corrosion-resistant materials such as nylon, plastic, polyethylene and steel mesh are used to construct a cage.

Recently, corrosion-resistant, antimicrobial copper alloys (Copper-zinc brass, copper-nickel and copper-silicon) are used in cage construction to prevent biofouling. It prevents adhesion and growth of disease-causing pathogens to the cage, and thus minimises the chances of infection. Cages can be either suspended in freshwater by a buoyant collar, (wood, bamboo, plastic pipe) and floats (Styrofoam or oil drums), or can be completely submerged in the ocean. The size of cages may vary from about 30 or 40 ft^3 to several hundred cubic feet.

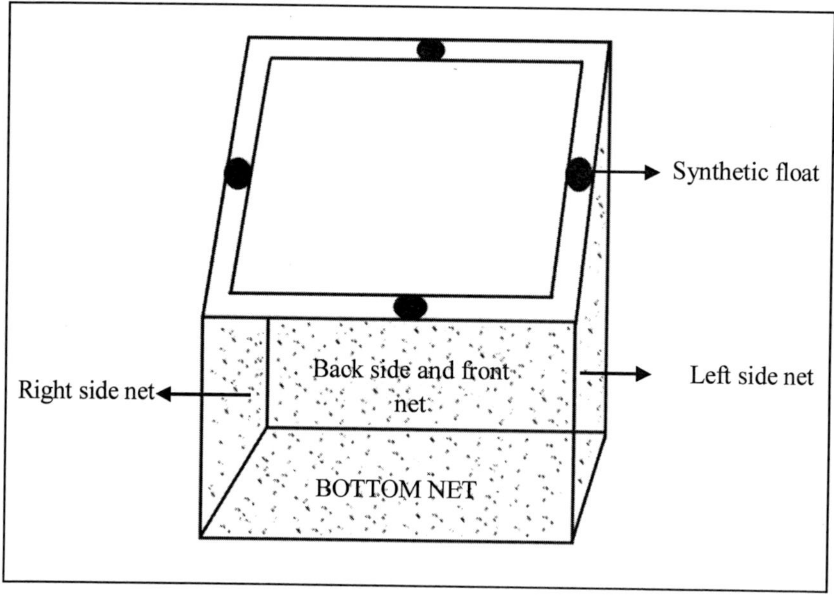

Fig. 4.3: A model cage for fish culture (enclosed on all sides except the top). Some cages are enclosed from all sides (submerged type).

4.3.2 Pen culture system

In pen culture system, the bottom of the enclosure is formed by the bed of the water body, while the other sides are enclosed by net fencing (Fig. 4.4). A pen can be constructed with frameworks of wood, bamboo, metal, etc., on which net fencing is done to form an enclosure. The net fencing is fixed to the bed of the water body by using pegs or sand bags to prevent the escape of the fish. The mesh size of the net should be small enough to prevent the escape of fry and fingerlings from the pen.

Pen culture system should be constructed in a zone of the water body where water flow is sufficient enough for replenishment of oxygen and flashing away of waste. Pen culture system is generally constructed in the intertidal, sublittoral and seabed zones in marine and brackish waters, while in the case of shallow freshwater bodies it is constructed in areas closer to the shore.

Shape and size of the pens: The shape of pens may be circular, square, and rectangular. The overall size of pens depends mainly on location and water depth, while the height of the pens is selected on the basis water level during entire culture periods and the jumping behaviour of cultured fish species.

Types of pen: On the basis of fencing materials used, pen culture system can be categorised into the following two types (http://aquafind.com/):

1) **Bamboo screen pens:** It is the simplest type of pen in which net fencing is done around the bamboo poles fixed to the bed of water bodies. This type of pen is generally constructed in narrow and shallow rivers, flooded fields, etc.
2) **Monofilament cloth fencing pens:** This type of pen is surrounded by monofilament netting material. The screen wall is arranged just like a fry net.

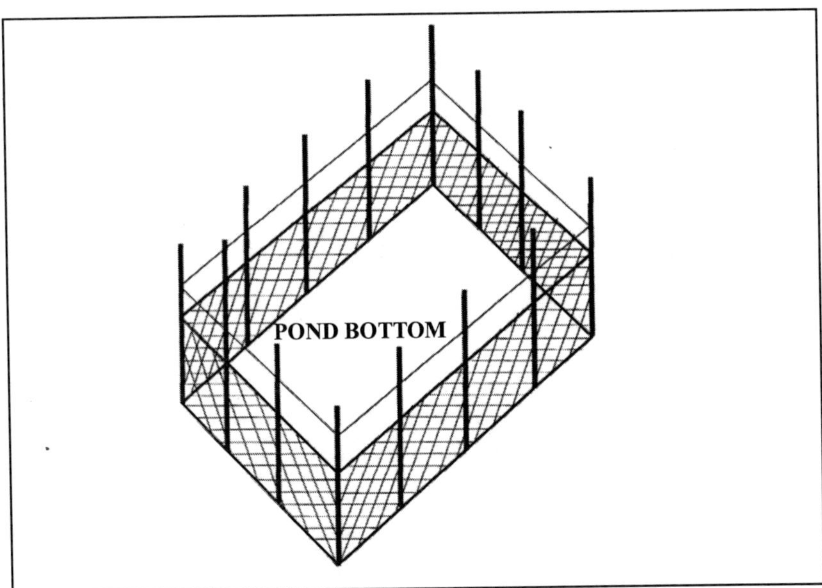

Fig. 4.4: A model pen for fish culture (enclosed on all sides except the top and bottom).

4.3.3 Advantages and disadvantages of pen and cage culture

Both cage and pen culture system can be set into any suitable water body. Though there are differences in their construction, both cage and pen culture systems have some common advantages and disadvantages when compared to other traditional open culture method.

Advantages:

1) **Ease of harvest:** Harvesting of fish from cage and pen is much easier than in the natural waters. Fish can be harvested without draining water.
2) **Minimal risk of predators:** Culturing of fish in an enclosed area as carried out in pen and cage culture safe them from their natural predators (Kumar and Karnatak 2014).
3) **Low initial investment:** Compared to the cost associated with pond construction and its related infrastructure, both pen and cage culture in an existing body of water is less expensive.
4) **Adequate environment:** As in both cage and pen culture system favourable environmental conditions are maintained along with artificial feeding, any varieties of species can be cultured.
5) **Resource flexibility:** It is possible to exploit underused water body to produce fish as any suitable water body (lakes, ponds, mining pits, streams, rivers, and sea) can be used to set up a cage or pen culture system.
6) **Disease control:** Fishes (behaviour, symptoms of infection) can be more easily observed in the enclosure culture system. Recently, antimicrobial copper alloys are used in cage culture to prevent biofouling.

Disadvantages:

1) **Dependence on artificial feed:** Cost of production in cage or pen is more compared to the traditional method, as fish depend completely on an artificial complete diet.
2) **Risk of theft:** Caged fish are often a target for poachers and vandals as fishes can be easily harvested from cage than in natural waters.
3) **Predation:** Predators like turtles, snakes, otters, raccoons often damage the cage and eat the reared fishes.

4) **Diseases:** Due to over-crowding in the cage, once a single fish get infected, it spreads rapidly among the fish. In addition, parasitic diseases are often transmitted from the surrounding wild fish.
5) **Limited production:** Compared to open pond culture systems, overall production in a cage is low, mostly because of the stress associated with high fish densities, high incidence of disease, low dissolved oxygen content and increase ammonia concentration in and around the cage. However, maintaining proper water flow (natural or artificial) through the cage minimises this problem to some extent (Kumar and Karnatak 2014).
6) **Pollution:** Since cage culture involves high stocking density in an enclosed area, depletion of dissolved oxygen, accumulation of ammonia and other excreted materials are common problems in cage particularly when there is no water flow through the cage.

Advantages over each other: Beside this common advantages and disadvantages, both cage and pen culture systems have some advantages over each other.

1) The towability or mobility of the cage is the most specific advantage of a cage over the pen culture system. For example, harvesting is much easier from a cage culture system than from a pen culture system.
2) As pens are generally larger than the cages, construction costs of pens are more than that of the cages.
3) Unlike cage, the bottom of the pen is the bed of the water body. This is advantageous over cage fish in pen are able to procure natural food and exchange of materials with the bottom.

4.4 COMPOSITE FISH CULTURE OR POLYCULTURE

The culture of fast growing compatible species of fish of different feeding habits togetherly in the same pond in order to obtain high production of fish per hectare of water is called mixed fish farming or composite fish culture or polyculture.

Principle: Culturing a single type of fish is not profitable because it cannot utilise all the natural food available at all the water levels of a pond (Fig. 4.5). Thus, in monoculture systems, many niches are unoccupied by fish and available food of these niches wasted. The principle behind polyculture is the exploitation all available niches of a pond by culturing fast growing, non-predatory, and non-compatible fishes of different feeding habits. Polyculture is beneficial to an aquaculturist as:

1) There is no competition for space between the compatible species as they occupy different ecological niches in a pond.
2) There is no competition for food between the compatible species as they have different food habitats (Fig. 4.6).
3) It is a scientific method based on ecological microhabitat or niche which results in high production with low investment cost.

Thus, employing polyculture an entire pond and all the food available in the different zones of the pond can be used profitably and sustainably.

Species interrelationship in polyculture: In India, mixed fish culture practice was started with three Indian major carps (IMCs) namely catla, rohu and mrigal. When, catla, rohu and mrigal are cultured in the ratio of 3:3:4 or 4:3:3 a good yield from a pond can be obtained as they have different food habitats (Alikunhi 1957; Chakrabarty et al. 1979).

1) Surface feeder: Catla is a surface feeder, feeds mainly on zooplanktons.
2) Bottom feeder: Mrigal feeds on decaying vegetation and epiphytic planktons from the bottom of the pond.
3) Mid-column feeder: Rohu is a mid-column feeder eating mainly decaying macro-vegetation, filamentous algae, periphyton, etc.

Along with IMCs, cultivation of compatible exotic carps such as silver carp, grass carp and common carp in the same pond gives a more profitable outcome. Of these three exotic carps (Table 4.2):

1) Silver carp is a surface feeder, but it does not compete with catla as it feeds on phytoplankton, while catla feeds on zooplanktons.
2) The omnivorous bottom feeder common carp, feeds mainly on insects, crustaceans, crawfish, benthic worms, and aquatic plants, while mrigal feeds on decaying vegetation and epiphytic planktons.
3) The grass carp feeds higher aquatic plants and submerged terrestrial vegetation and occupy a niche not utilise by silver carp, common carp and IMCs.

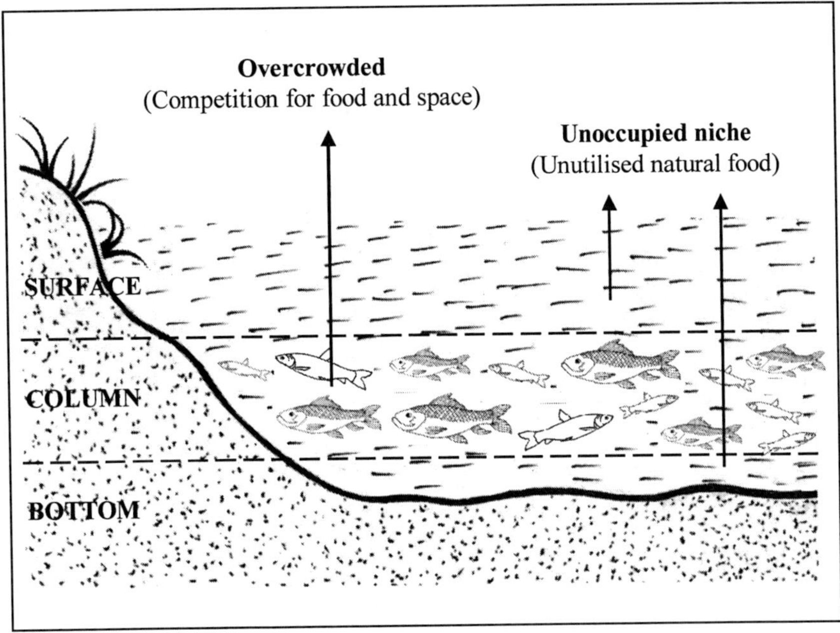

Fig. 4.5: Nonscientific pond fish culture showing unutilisation of natural food available at different water level.

Beside this, the association of grass carp in polyculture has an indirect benefit also. The semi-digested excreta of grass carp serve as the food of bottom dwellers like mrigal and common carp. However, grass carp feeds on supplementary feeds like rice polish and oilcakes. Thus, to avoid competition among grass carp, rohu, mrigal and common carp for taking supplementary feeds, the pond should be supplied with aquatic weeds also for grass carp.

Table 4.2: Types in preferable species for polyculture, their ecological niche and feeding habits.

Species	Feeding zone	Feeding habits
Catla (*Catla catla*)	Surface feeder	Zooplankton feeder
Rohu (*Labeo rohita*)	Column feeder	Omnivorous
Mrigal (*Cirrhinus mrigala*)	Bottom feeder	Detritivorous
Silver carp (*Hypophthalmichthys molitrix*)	Surface feeder	Phytoplankton feeder
Grass carp (*Ctenopharyngodon idella*)	Surface, column and marginal feeder	Herbivorous
Common carp (*Cyprinus carpio*)	Bottom feeder	Detritivorous or omnivorous

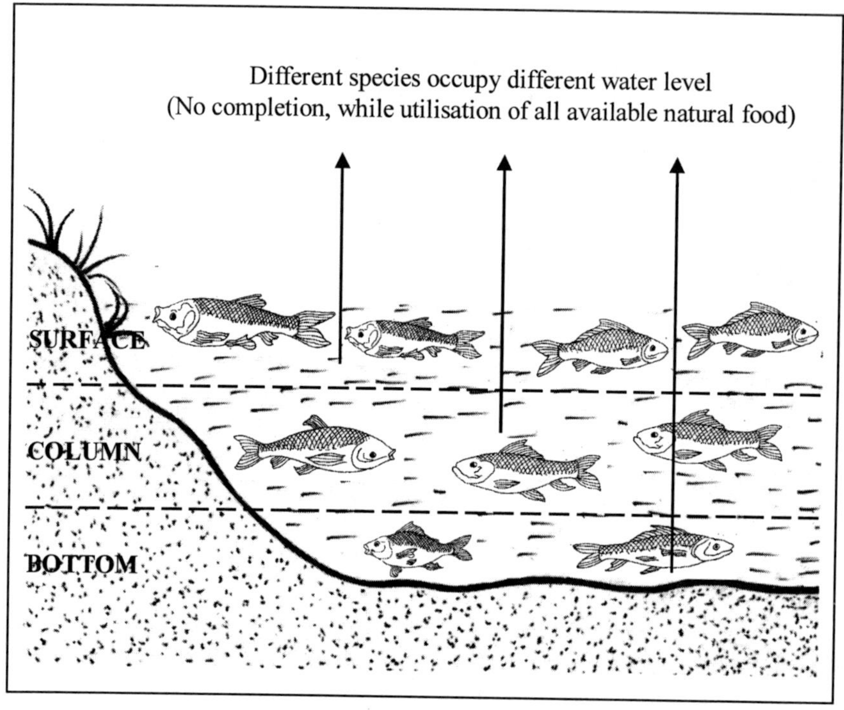

Fig. 4.6: Composite fish culture method showing the utilisation of natural food available at different water level.

Outcome of polyculture: Polyculture results in high production which is generally 5-6 times more than the general production. A number of experiments have been conducted using different combinations of IMCs and exotic carps in between 1962 to 1963.

1) Culture of IMCs alone yield production of around 2088 kg per hectare per year.
2) When only exotic carps were cultured the average yield of 2900 kg per hectare per year was obtained.
3) When a combination of IMCs and exotic carps were cultured the average yield was 3000–3500 kg per hectare per year (Alikunhi et al. 1971). Thus, a composite fish culture of six species comprising of three Indian major carps and three exotic carps is more profitable.
4) As per ICAR data, the production yield of 3448-5894 kg/ha was obtained in six months and a production yield of 6191–7332 kg/ha was obtained in 8 months in the tanks of Karnal in Haryana (Sukumaran, 1976).

Significance: Human population is growing day by day and simultaneously aquatic habitats are also destroyed bit by bit. The production of enough protein-rich, nutritious food sustainably to feed everyone is one of the major challenges in the present day. In this direction, production of fish through polyculture is much more efficient compared to traditional method of fish farming. For example, the culture of IMCs alone yielded production of around 2088 kg per hectare per year. This will manage the present crises of getting animal protein from the existing water resources up to some extent.

4.5 BROODSTOCK MANAGEMENT

For sustainable commercial aquaculture production, proper control of reproductive function in captivity is very essential. Broodstock refers to a group of sexually mature individuals of a cultured fish species used for breeding purposes.

Broodstock management involves all the appropriate measures taken to a captive fish group for maximum survival, enhance gonadal development and increase fecundity. Thus, broodstock management is related to all the segments of aquaculture production cycle and is briefly discussed below:

1) **Broodfish selection:** The broodstock must be selected early from the fingerlings available rather than from the commercial stock. For example, optimum gonadal development is obtained when common carp fry produced in February-March are selected as broodstock. The broodfish must possess improved qualities such as rapid growth potential and ability to tolerate adverse conditions like dissolved oxygen deficiency and adverse water quality, strong appetite, omnivorous feeding regime. Following are the key morphological features which can be used during selection of good quality broodfish
    a) It should be in good health.
    b) The fish should be free of body wounds or injury.
    c) No parasitic infection.
    d) No fin or body deformation.
    e) The body should possess the required shape and proportions.

2) **Pond preparation:** Optimum oxygen concentration, temperature, and pH of the water favours good gonad development. Thus, it is necessary to manage the water quality of the broodstock pond before stocking.
   a) If pH is less than 6.5 or alkalinity is less than 75 mg of calcium carbonate, liming is recommended before stocking. This will increase the pH of soil and pond water which causes the release of nutrients into the pond water and also neutralise toxic substances (Fig. 4.7).
   b) Manure (cow, chicken, duck, pig manure) rich in nitrogen and phosphorus, compost at the rate 25-50Kg/ 100 m^2 should be applied few days after liming.
   c) Water filling of a pond is generally done in two stages: in first stage 20-30 cm of the pond is filled with water which allows growth of natural food. In the second stage complete filling is done after one week.
3) **Stocking density:** Generally 0.5-1 train size and 1-2m deep is the most preferable size of the broodstock pond. Before stocking, the dykes should be protected by vegetation to prevent erosion or flooding during rainy season. Both inlet and outlet of the pond must be provided with water control structures and net to prevent entry of unwanted fish and escape of broodstock. In general, the stocking density varies from 100 to 300 per hectare.
4) **Stocking with carnivorous fish:** To control the wild fish which might enter the broodstock pond and compete for food and space with the broodstock, many aquaculturists prefer stocking of broodstock with smaller carnivorous fish (~ 200-400 individual per hectare).
5) **Broodstock nutrition and feeding:** Gonadal development, fecundity, egg and sperm quality, seed productions are dependent on essential dietary nutrients. It has been found that broodstock fish feeding with vitamin E, ascorbic acid, lipid and fatty acid (Highly unsaturated fatty acids) rich food have high fertilisation, fecundity and egg quality (Izquierdo et al. 2001). Cuttlefish, squid and krill meals are also important broodstock diets. However, it is recommended to reduce the feeding during the cool season. For efficient feeding, the time (2 times morning and evening) and feeding place of the pond should be same every day.
6) **Fertilisation:** Even after pond preparation, fertilisation of the broodstock pond is necessary for maintaining a continuous production

of natural feed. However, excess fertilisation often results in fish stress as it unstabilises the water quality. A simple method of knowing when to fertiliser is to examine water clarity. Light penetration can be measured using a Secchi disk. If the sunlight penetrates the water 18 inches or more, a fertilisation program should be implemented. Depending on the type of manure (buffalo, cattle, pig, duck, poultry, etc.) fertilisation rate varies from 1.5-4.5Kg/ 100m^2. Instead of manure, compost can also be used at a rate of 25 kg/100 m^2 in a week.

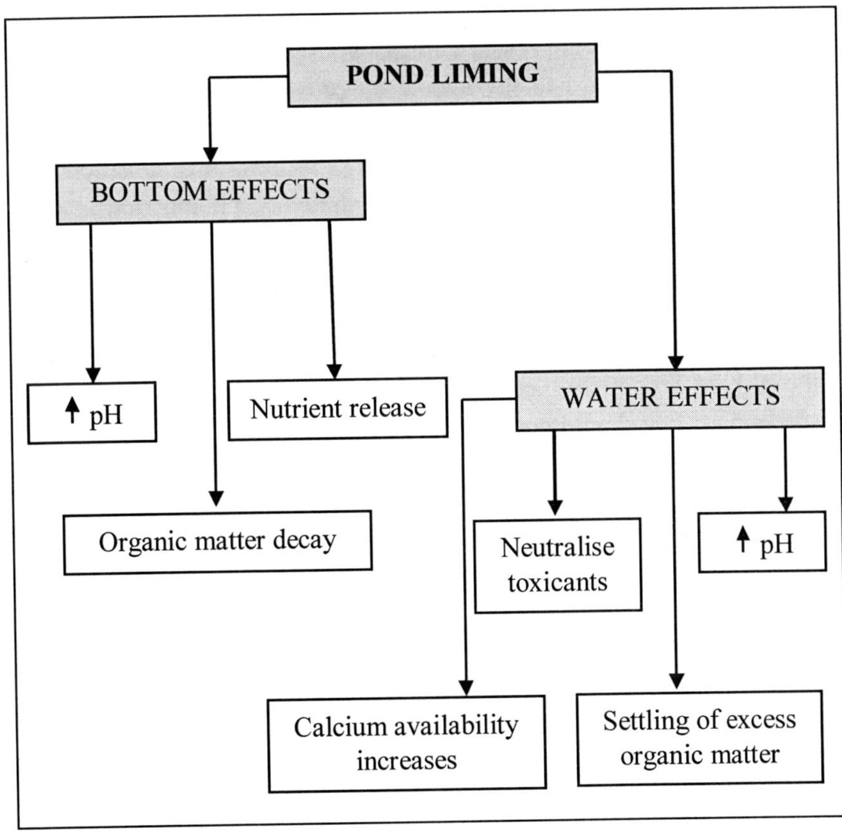

Fig. 4.7: Positive effects of liming on pond bottom and pond water.

7) **Sex separation:** Control of reproductive function in captivity is essential for commercial aquaculture. Thus, accurate identification of male and female population and maturity stage is very essential. For example, the body of female carp is plump and its genital opening is situated above the genital papilla. On the other hand, the body of male

crap is slender and its genital opening is found behind the genital papilla. On maturity both male and female show unique external signs:

a) Mature male: 1) it generally releases milt under a slight abdominal pressure, 2) its belly is not blown up, but rather slim and 3) it has callosities on the head.

b) Mature female: 1) it has a well-rounded and soft or semi-soft belly, 2) its genital papilla is erect and reddish and 3) its anal opening is enlarged and protruding.

Some aquaculturist prefers two sets of ponds for ripening males and females separately and two sets of ponds for the spent spawners. This is advantageous as it will prevent wild spawning and also help the farmer in taking special care to the female by providing protein rich foods (FAO, 1985).

8) **Spawning:** After selection of mature broods, manipulation of environmental factors like photoperiod (shortened photoperiod), water temperature (high) can be used for artificial spawning when fry are desired. In addition to this hypophysation (induced breeding) can be used to induce fish to spawn.

9) **Post-spawning maintenance:** Following spawning fish, particularly the female broodfish should be transferred to a resting pond after spawning. They should not be mixed with broodstock fish. During this period, they must be fed with an energy rich diet like rice, cassava, potatoes, etc. This will help in recovering the weight loss occurs during reproduction and also keep them stress-free.

4.6 PREPARATION AND MAINTENANCE OF FISH AQUARIUM

4.6.1 Fish aquarium

Fish aquarium (“aqua” meaning water and “arium” meaning a place for relating to) is an artificially constructed tank or pool for keeping and displaying ornamental fish. The term aquarium was coined by English naturalist Philip Henry Gosse. The concept of fish keeping in an aquarium was fully developed in 1850 by the chemist Robert Warington.

They are typically made up of glass or high-strength acrylic. Most of the fish aquariums are constructed for ornamental, research and breeding purposes. Some of the most common examples of freshwater home aquarium fish species are: *Carassius auratus*, *Channa spp.*, *Nandus nandus*, *Amblypharyngodon mola*, *Puntius spp.*, *Acanthopthalmus pangia*, *Acanthocobitis botia*, *Botia rostrata*, *Botia dario*, *Ailia coila*, *Balitora brucei*, *Barilius barna*, *Batasio tengana*, *Erethistes pussilus*, *Badis badis* etc.

4.6.2 Construction of fish aquarium

An aquarium is generally made for decorative purposes in offices, hotels, shops, drawing room, etc., public exhibition, research purposes and breeding purposes. During the construction of a good aquarium following facts must be considered (Ngueku 2014):

1) **Size:** The size of the aquarium depends upon the purpose and available area where it is being installed. The size of a standard rectangular shaped home aquarium is 60 × 30 × 38 cm (18 gallons). However, large size (many thousands of gallons) aquariums are also available which are mainly constructed for public exhibition. For example, an aquarium (27 ×108 feet) of Dubai Mall in Dubai has a capacity to hold 10 million litres (2, 64 million gallons) of water. The aquarium has more than 33,000 living animals, including over 400 sharks and rays.
2) **Shape:** The shape of an aquarium can be rectangular, square, triangle, column, pyramid, etc. However, most of the aquariums are generally rectangular shaped.
3) **Material:** The materials, mostly used for aquarium construction are glass, high-strength acrylic and silicone sealant.
4) **Components**: The major accessories of a hobbyist fish aquarium include the following:
    a) **Composts:** It includes gravels, sand, rocks, floating and rooted plants, bog-wood, small artificial caves, etc. They provide an artificially favourable environment for the growth of plants and survival of tank inhabitants.
    b) **Artificial lighting system:** A light bulb of about 25w (240v) is vital to the health of aquarium fish and plants. It helps hobbyists to observe aquarium inhabitants and provides vital energy to photosynthetic plants, anemones, or corals. For fish tank, 3 watts of bulb per gallon of water is generally recommended.

c) **Filtration system:** It removes chemical waste products like ammonia (excretory product of fish) from aquaria.
   - In biological filtration, bacteria are used to convert ammonia into slightly less toxic nitrites and then to less toxic nitrates.
   - Mechanical filtration removes particulate material like uneaten food, faeces or plant or algal debris from the water column. Example- magic jet power filter.
   - Chemical filtration removes of dissolved wastes from the aquarium. Example- activated carbon, foam fractionation, etc.

d) **Aquarium heater and chiller:** Aquarium heaters raise the water temperature to a desired level suitable for the fish. For example, in a tropical aquarium, the temperature should be maintained within the range of 65 °F-85^{0}F (18-29^{0}C) depending upon the species.

   Sometimes accessories like large UV sterilisers, metal halide lights or water pumps may heat the water to a temperature higher than optimum level. In this situation, a chiller may be necessary to bring back the temperature to the optimum range. Example- Prime drop-in chiller, Arctica titanium chiller, etc.

e) **Aquarium hood:** It is made up of a plain glass sheet, plywood or metal-hood. It prevents fishes from jumping out, keeps out dust, and reduces heat loss and evaporation. It also holds light bulbs or tube.

f) **Air pumps:** It increases water circulation and supply adequate gaseous (water-air) exchange. It is made up of motor pump, air tubes and joints, controller and a power source. The end of the tube submerged in water is often fitted with an airstone or aquarium bubbler. It increases the rate of water circulation and gas exchange by generating steam of small bubbles.

g) **Thermometer:** Some aquariums are also fitted with a thermometer to get an accurate temperature reading. Example- glass alcohol thermometers, adhesive external plastic strip thermometers, and battery-powered LCD thermometers.

4.6.3 Maintenance of fish aquarium

Regular maintenance like water changes and cleaning of the aquarium are vital for the survival of aquarium inhabitants. Following are the important aquarium maintenance practices one can follow to keep the healthy aquatic environment and thriving fish for a long period of time.

1) **Water changes:** The most serious problem arises in a fish tank is the accumulation of waste (e.g., ammonia, uneaten food, etc.). At least 50-percent of the water change should be performed once a week. A good water changing method involves vacuuming of gravel, which will eliminate uneaten foods and other residues that accumulated on the aquarium bottom.
2) **Quality checking of replacement water:** It is important to check the quality of replacement water before its addition to the aquarium as most tap water contains either chlorine or chloramines. If it is rich in chlorine or chloramines, a water conditioner can be used to neutralise the chlorine.
3) **Regular testing aquarium water:** It is vital to check water parameters such as pH, temperature, nitrates, nitrites, and carbonate hardness on a regular basis. Thus, an aquarium must be provided with a thermometer and a pH meter.
   - Optimum pH range for most species is 6.5 – 7.5.
   - KH (carbonate hardness) should be kept above 4.5 dH or 80 ppm. In extreme cases KH can be raised by about 1 dH (17.8 ppm) by adding half a teaspoon of baking soda per twenty-five gallons of water.
   - Nitrites should be undetectable at all times.
   - Nitrates should be kept below 10 ppm in freshwater fishes.
   - In tropical aquarium, the temperature should be maintained within the range of 65^0F-85^0F (18-29^0C)
4) **Glass cleaning:** Algae are important for the aquarium, but the attachment of algae on the glass makes it difficult for hobbyists to observe the fish. Thus, it is important to clean algae from the glass once a week using a scraper made of metal or plastic blades or an abrasive pad.
5) **Vacuuming:** Uneaten food and faeces often settle at the bottom of the aquarium. Thus, a good vacuuming at least once a week is highly recommended to remove this waste. A siphon with a gravel tube at one end can be used to clean the gravels.
6) **Filter Cleaning:** A mooth flow of water is often impaired due to clogging of biofilter by debris. Thus, filter inserts (floss, algone, activated carbon) should be changed at least every four weeks. It is

also necessary to clean the filter by during the water change at least once a month.

7) **Avoid overfeeding:** Excess feeding is not only unhealthy for the fish, but also makes tank dirty, clog filter, etc. Quantity and frequency of feeding depend on several factors like size and age of fish, water temperature and the quality of diet. For example, younger fish require higher feeding rates than older fish. Cyprinids (goldfish, koi, barbs, danios) require more frequent feedings as they lack a true stomach. Thus, it is important for a hobbyist to develop own feeding chart through regular observation of the fish behaviour in the aquarium.
8) **Disease control:** Aquarium fish are often subjected to viral, bacterial and fungal diseases like fin rot, swim bladder disorder, dropsy, etc. All unhealthy fishes noticed with disease signs should be collected and treated with malachite green, formalin baths, copper sulphate and benzalkonium chloride dips, oral methylene blue and antibiotics in a quarantine tank.
9) **Additional maintenance:** Other good aquarium maintenance practices include stocking with healthy and disease free fish, proper aquarium lighting, cover glass cleaning, regular observation of fish behaviour, etc.

4.6.4 Importance of aquarium

1) The aquarium and aquarium fish is a source of employment and income for individuals/companies of different countries associated with its design, construction and supply.
2) Some aquarium fish helps in the control of intermediate host (e.g., mosquitoes) of many parasitic diseases like malaria.
3) It is used for decorative purposes as it improves the scenic beauty of the environment. Aquarium fish like medaka and zebrafish are widely used as a fish model for studying human disease. Similarly killifish, toadfish and damselfish are used as a fish model for ageing, hepatic encephalopathy, sickle cell anaemia, cancers, etc. (Schartl 2014).
4) Watching fish aquarium has a number of health benefits like it reduces stress, lowers blood pressure, reduces hyperactivity disorder in children; control Alzheimer's disease, etc.
5) The fish aquarium is widely used in "Pet therapy" to reduce loneliness, anger, depression, and stress.

4.7 ROLE OF WATER QUALITY IN AQUACULTURE

The water quality has significant impacts on overall fish farming or pisciculture. Optimal water quality, including physical, chemical and biological must be monitored for proper growth and survival of fish or more correctly for optimum fish production. The different parameters, their acceptable range and the amount causing stress in fish are highlighted in the below table (Table 4.3).

Table 4.3: Suggested water-quality parameters for pond water fishery (Bhatnagar and Devi 2013).

Sr. no.	Parameter	Acceptable range	Desirable range	Stress
1	Temperature (^{0}C)	15-35	20-30	< 12, > 35
2	Water colour	Pale to light green	Light green to light brown	Clear water, dark green and brown
3	DO (mg L^{-1})	3-5	5	< 5, > 8
4	BOD (mg L^{-1})	3-6	1-2	> 10
5	CO_2 (mg L^{-1})	0-10	<5, 5-8	> 12
6	Alkalinity (mg L^{-1})	50-200	25-100	< 20, > 300
7	Hardness (mg L^{-1})	> 20	75-150	< 20, > 300
8	Ammonia (mg L^{-1})	0-0.05	0 - < 0.025	> 0.3
9	Nitrite (mg L^{-1})	0.02-2	< 0.02	> 0.2
10	Nitrate (mg L^{-1})	0-100	0.1-4.5	< 0.01, > 100
11	Phosphorus (mg L^{-1})	0.03-2	0.01-3	> 3
12	Plankton (No. L^{-1})	2000-6000	3000-4500	< 3000, > 7000

Physicochemical and biological parameters like temperature, transparency, turbidity, water colour, carbon dioxide, pH, alkalinity, hardness, unionized ammonia, nitrite, nitrate, primary productivity, BOD, plankton population, etc. determine the quality of water (Table 4.4).

Table 4.4: Influence of different physicochemical characteristics of water on fish and their control measures.

Sr. No.	Parameter	Reason and influence (effects)	Optimum level	Management or control
1	**Dissolved oxygen**	Photosynthesis of phytoplankton is the major contributor of DO. However, when plankton density is high, it limits the penetration of sunlight in water, and thus reduces the photosynthetic oxygen production. Moreover, the solubility of oxygen decreases as the temperature increases. Oxygen has a direct effect on feed consumption and metabolism and indirect effect on environmental conditions. Low levels of dissolved oxygen can cause changes in oxidation state of substances from the oxidised to the toxic reduced form.	In and around 5 ppm saturation level or should be above 4 mg/l	Aeration is a proven technique for improving DO availability. Any sort of agitation improves the DO content and among which paddle-wheel, aerators aspirators are most common.
2	**Temperature**	Optimum temperature is necessary for metabolism, active behaviour and growth rate.	For cold water fishes: 14^0C-18^0C For warm water fishes 24^0C-30^0C	Use of aerator helps in breaking thermal stratification. Planting of trees gives shade and maintains adequate temperature.
3	**Turbidity**	Turbidity limits light penetration, thereby limiting	Optimum	It can be controlled by

		photosynthesis in the bottom layer. Higher turbidity can cause temperature and DO stratification in ponds. High turbid water can cause clogging of gills or direct injury to tissues of fishes.	visibility ranges from 40-60 cm	application of organic manure at 500-1000 kg/ha, gypsum at 250-500 kg/ha or alum at 25-50 kg/ha.
4	**Ammonia**	Fish excrete ammonia. Thus, in highly stocked fish ponds, ammonia concentrations can become high. The excess protein diet also increases the ammonia level in the water. This directly affects the fish and has a number of harmful side effects like increased disease susceptibility, suffocation, organ failure.	0.02-0.05 ppm	Addition of salt at the rate of 1200-1800 kg/ha reduces toxicity. A biological filter may be used to treat water for converting ammonia to nitrate. This nitrate can be then converted to harmless nitrate through nitrification process.
5	**Hydrogen sulphide**	The fresh water fish pond should be free from hydrogen sulphide because at a concentration of 0.01 ppm fish lose their equilibrium.	Below 0.01 ppm	Frequent exchange and increase of pH through liming can reduce its toxicity
6	pH	Water pH affects fish metabolism, physiological process, the toxicity of ammonia, hydrogen sulphides and solubility of nutrient and thereby well-being and fertility of fish.	6-9	pH can be increased by the application of lime, agriculture gypsum

7	**Total hardness**	Numerous inorganic (mineral) substances are dissolved in water. Among these, the metals calcium and magnesium, along with their counter ion carbonate (CO_3^{2-}) are the main. Hard water has the capability of buffering the effects of heavy metals such as copper or zinc which are in general toxic to fish.	Greater than 40-400 ppm	Low hardness can be treated with lime,
8	**Carbon dioxide**	In highly stocked fish ponds, CO_2 concentrations can become high as a result of respiration. The CO_2 reacts with water to produce H_2CO_3. It ionises to release H^+. Thus, high CO_2 decreases pH drastically.	Less than 5 ppm	Carbon dioxide can be removed by chemical treatment of pond water with liming agents such as quicklime, hydrated lime or sodium carbonate.
9	**BOD and COD**	This is an indication of both sewage and industrial pollution. High levels of BOD and COD can deplete the oxygen in water	BOD- less than 10 mg/l COD- less than 50 mg/l	Hydrogen peroxide (H_2O_2) has been used to reduce the BOD and COD. Controlling wastewater discharge into water bodies.
10	**Nitrate**	Excess nitrate is toxic to aquatic animals as it helps in the explosive growth of plants and algae called eutrophication.	0.1 to 4.5 mg/liter	Reducing the flow of runways water from agricultural fields, tea

				gardens, industries, households, etc.
11	**Phosphate**	Similar to nitrate, excess phosphate toxic to aquatic animals as it helps in the explosive growth of plants and algae called eutrophication.	0.2- to 0.3 mg/ liter	Reducing the flow of runways water from agricultural fields, tea gardens, etc.
12	**Nitrite (NO^{2-})**	Excess nitrite oxidises haemoglobin to methemoglobin in the blood. It hinders respiration, damage nervous system, liver, spleen and kidneys of the fish.	Desirable range is 0-1 mg/L. It should not exceed 0.2 mg/L in freshwater and 0.125 mg/L in sea water	Reduction of stocking densities, biological filtration, addition of small amounts of certain chloride salts, use of biofertilisers (Bhatnagar and Devi 2013)

4.8 SEWAGE-FED FISHERIES

Sewage is the liquid waste (solution or suspension) discharged from the domestic and industrial sources within an area. Utilisation of sewage for fertilising fish ponds is called sewage fed fisheries. A sewage fed pond does not require additional fertilisers and supplemented food, and thus it reduces the cost of fish culture.

Composition of sewage: Chemically it consists of approximately 99% water and 1% inorganic and organic matter either in suspended and soluble forms. Phosphorus and nitrogen are the main constituent of sewage. It also contains traces of inorganic compound like zinc, copper, ammonium, nitrate, etc. and organic compound like lignocellulose, cellulose, proteins, fats, etc. It is also rich in detergents and microbes like virus, bacteria, fungi and protozoa. It has high conductivity, high alkalinity with a pH in between 7 and 8.

General method adopted in sewage-fed fisheries: Sewage has low oxygen content (DO), while high ammonia, BOD, CO_2, sulphur and bacterial load. As this compound has detrimental effects on the fish life, it is essential or recommended to treat the sewage before releasing it to a fish pond. The general step used in sewage-fed fisheries involves:

1) **Treatment of sewage:** To kill pathogenic microorganisms, prevent anoxia, reduce organic content, sewage is first treated by following three methods:

   a) **Mechanical:** It consists of screening, filtration and sedimentation to remove suspended materials and greasy and oily materials. In screening, sewage is passed through a series of filters with different pore sizes. It works like a sieve to remove floating materials like plant leaves, wood pieces, plastic bags, etc.

   Sedimentation is done by passing the sewage into a tank or basin by high velocity. When sewage enters the tank there is a sudden drop in velocity and this helps in the sedimentation of coarse solid materials (sand, ash and others). Sometimes chemical coagulants like alum, iron salts, lime, soda, activated silica and polyelectrolytes are added to facilitate sedimentation of colloidal waste.

   Greasy and oily materials (fat, oils, soaps, wax) are removed by skimming. During this process, oily and greasy materials coagulate and solidify on the water surface, which then can be removed either manually or mechanically

b) **Chemical:** Even after biological treatment, sewage still contains nitrogen and phosphorus salts which can cause serious eutrophication in aquatic ecosystems. During chemical treatment, phosphorus salts are precipitated by liming, while nitrogen present in the form ammonia is removed by volatilization at a high pH. Finally, chlorination using either sodium or calcium hypochlorite or chlorine is generally done for disinfection.

c) **Biological:** In this step microorganisms (fungi, bacteria, protozoa, algae, etc.) are used to decompose the unstable organic matter into simpler forms. The microbes break down organic matter to CO_2, H_2O, nitrogen, sulphates and other inorganic substances.

In addition, nitrogen can be removed by biological oxidation of nitrogen from ammonia to nitrate (nitrification), followed by denitrification (reduction of nitrate to nitrogen gas). Similarly, phosphorus can be removed biologically by polyphosphate accumulating organisms (PAOs), which can accumulate large quantities of phosphorus within their cells.

2) **Storage in the second sedimentation tank:** After ten days of initial treatment (mechanical, biological and chemical), sewage is then generally passed into another tank or basin called second sedimentation tank or waste stabilisation pond. In this tank, sewage is stored for about 15 to 20 days. During this period, sedimentation of remaining solid particles takes place; the sewage loses its foul odour and it has also become rich in plankton flora (algal bloom) and fauna.

3) **Dilution:** Even after the treatment of sewage, it cannot be directly released into water bodies mainly because of its low oxygen content, high ammonia, CO_2 and H_2S. Thus, before releasing the treated sewage water into nursery or stocking ponds, it must be diluted with fresh water. In general a dilution ratio of 1:5 or 1:4 used for dilution of sewage water with fresh water. Dilutions lower the level of $C0_2$, ammonia, CO_2 and H_2S below the lethal limits and also restore the DO level suitable for proper growth of planktons and fish development (Fig. 4.8).

4) **Release into the main pond:** It is generally recommended to load the nursery or stocking pond with treated and diluted sewage once a month. Loading to an extent of one-fourth or one-fifth of the water

level of the fish pond is found to be highly profitable. As sewage treated pond contains relatively more fertilisers and nutrients compared to non-sewage treated fish pond, fish production rate always is higher in sewage-fed pond.

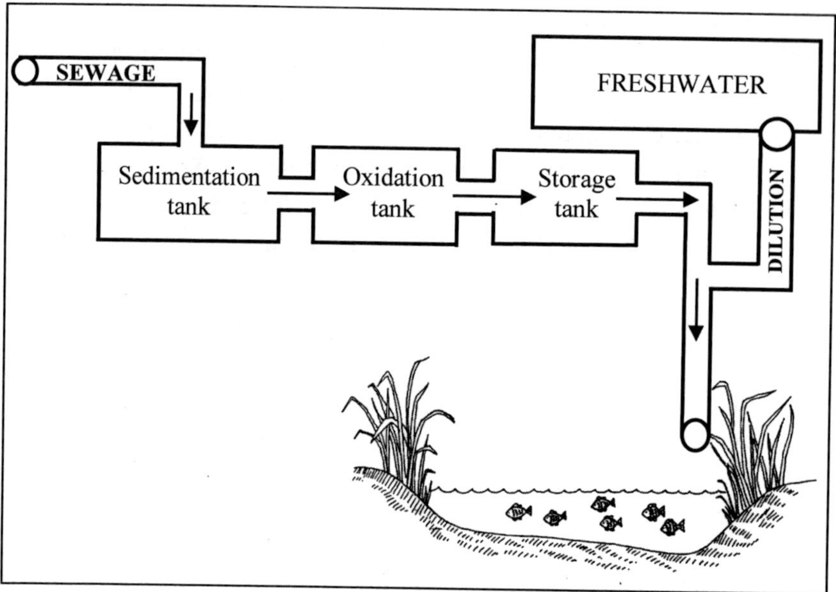

Fig. 4.8: Schematic representation of sewage-fed fishery technique.

Fish culture in sewage-fed ponds in India: Air breathing fishes are generally preferred for culturing in sewage treatment ponds as they can survive in water with lesser dissolved oxygen content. The fish like *Clarias batrachus*, *Heteropneustes fossalis*, *Channa spp.*, *Tilapia mossambicus* and grass carp are the species of choice to be considered for culture in sewage treated ponds. As carps are very sensitive to low dissolved oxygen (DO) content, and thus they must be raised in ponds receiving sewage water after proper dilution. Silver carp, catla, rohu, mrigal, common carp and grass carp in a ratio of 25: 15: 10: 25: 20: 5 is found to be highly profitable in sewage-fed fisheries. Rearing of Catla, rohu, mrigala, common carp and scale carp in a ratio of 40: 10: 20: 20: 10 also gives profitable results.

Stocking density is comparatively higher in sewage-fed pond than the normal fish ponds. For example, the stocking density of *Cirrhinus mrigala* can be 10,000 per hectare in sewage-fed water, while it is 5000 per hectare in fresh water ponds. A total production of 220 kg/hectare in a composite culture of *Tilapia* and *Clarias* has been obtained in a study.

Similarly, stocking of rohu, catla and mrigal (7.5 cm length) in a ratio of 1:1:1 @ 550 kg/ha give an annual fish production of 3237 kg/ha (Ghosh et al. 1985).

Advantages of sewage-fed fisheries: The following are the advantages of sewage-fed fisheries:

1) A sewage fed pond does not require additional fertilisers and supplemented food. Thus, it reduces the cost of fish culture and at the same time increases the production rate of fish.
2) Compared to non-sewage fed pond, a high-stocking density can be maintained in sewage-fed pond. For example, the recommended stocking density of *Cirrhinus mrigala* for sewage-fed pond is 10,000 per hectare, while it is 5000 per hectare in fresh water, non-sewage fed pond.
3) The use of sewage in aquaculture results in high yields.
4) Moreover, the use of sewage in fisheries economises fertiliser and feed costs, resulting in higher profits.

Disadvantages of sewage-fed fisheries: The following are the disadvantages of sewage fed fisheries:

1) Fishes are often infected with Copepod parasites due to high organic load in the pond.
2) Sudden fall in oxygen level due to unplanned heavy sewage loading often results in fish mortalities.
3) It requires a large area for the setting of the treatment plant of sewage.
4) Since a high stocking density in maintained in sewage-fed pond, there is always a high chance of rapid disease transmission.

4.9 INTEGRATED FISH FARMING

Integrated fish farming is a farming system where fish culture (pisciculture) is carried out along with other animal or plant in order to achieve a high yield of both. It is advantageous as livestock excreta can be used as a fertiliser for the fish pond. Moreover, an integrated system offers maximum utilisation of the farming space. There are various combinations of livestock and agricultural crop in integrated aquaculture. Some of this are-

1) Fish culture cum poultry rearing

2) Fish culture cum duck rearing
3) Fish culture cum pig rearing
4) Paddy cum fish culture

4.9.1 Fish culture cum poultry farming

Inland fish culture in association with poultry rearing for meat (broilers) or eggs (layers) is a compatible business as it provides an aquaculturist a ready-made source of manure to fertilise the fish farm. The simultaneous farming of fish along with poultry is called fish culture cum poultry farming.

Principle: Fish and poultry are mutually complementary. The fish provide the fish meal, a protein-rich poultry feed for poultry farm. On the other hand, the poultry manure is rich in nitrogen, phosphorus, and potassium. Thus, it provides manure to ponds and hence either directly or indirectly (growth of planktons, the natural fish food) promote fish production. It is estimated that a chicken can produce approximately 40g of excreta per day. Thus, fish-poultry farming reduces costs of fertilisers and feeds for fish culture, while maximises the benefits.

Technique: Fish cum poultry farming has now become a popular rural enterprise in different states of the India. The basic technique involves-

1) **Breed and species selection:** Breeds or strains like RIR (Rhode Island Red), Hitkari, Vanraja, WLH (White Leghorn), ISA Brown, Kuroiler etc. are generally raised for meat or eggs in fish cum poultry integrated farming system.

   Fish species of high market demand like Indian major carps (rohu, catla and mrigal) and exotic carps (silver carp, grass carp and common carp) are generally preferred for fish cum poultry farming system in India.

2) **House construction:** Two methods are generally adopted for recycling the poultry excreta into fish ponds.
   a) **Method 1:** In this method, poultry house is constructed in a suitable space away from the fish pond. The poultry excreta from the poultry farms are collected every morning and stored it in suitable places. From this stored place excreta is applied in the ponds at regular instalments. Generally, the poultry excreta is applied to the pond every morning after sunrise at the rate of 50 Kg/ha/ day.
   b) **Method 2:** In the alternative method, poultry house is constructed above the fish pond, partially covering it. Thus, poultry excreta

directly fall into the fish pond. It is advantageous over the first method as i) no extra land for poultry house is required and ii) no extra labour for transporting manure to the fish ponds.

3) **Maintenance of poultry house:** Intensive poultry farming system is used in fish- cum poultry farming. The birds are confined to the house entirely. The intensive system is further of two types- cage and deep litter system.

   a) **Deep litter system:** In deep litter system, the floor of the poultry house is covered with litter like dry organic material such as chopped straw, dry leaves, hay, groundnut shells, broken maize stalk, sawdust, etc. up to a depth of about 8-12 inches. The birds are then reared over this litter. In about 10 months time it becomes fully built up poultry litter which can be used as fertiliser for the fish pond.

   b) **Cage system:** In this method, 3 to 8 hens are reared in small cages, usually made of metal. The main advantage of this system is that the greater number of birds is reared per unit of area.

4) **Stocking density:** In a pond of 1000m , around, a stocking rate of 600-1000 fingerlings of Indian major carps, (rohu, catla and mrigal) and exotic carps (silver carp, grass carp and common carp) is preferable. The species ratio consists of 40% surface feeders, 30% bottom feeders, 20% column feeder and 10-20 % weed feeders like grass carp. Generally, 30-50 broilers or chicken layers are adequate to fertilise a $1000m^2$ pond or about 500-600 poultry birds are enough for 1 ha of fish ponds (Gupta and Noble, 2001).

5) **Harvesting:** Depending on the demand of the fish in the local markets, partial harvesting of table size fish can be done 6-7 months after stocking. However, final harvesting is done after 12 months of rearing. Fish yield ranges from 3500-4000 Kg/ha/yr when 6 species are cultured and 2000-2600 Kg/ha/yr when 3 species cultured.

   Eggs can be collected daily in the morning and evening. A healthy bird lays about 200eggs in a year. However, as the egg laying capacity of these birds decreases after 18 months they must be sold after that period.

Benefits of fish cum poultry farming: The following are the advantages of integrated fish and poultry farming-

1) Poultry droppings are rich in nitrogen, phosphorus, potassium which promotes the growth of natural fish food like phytoplankton and

zooplankton. Thus, no additional cost is required for fertilisers and feeds of fish.

2) Fish cum poultry integrated farming system results in high production of fish, poultry eggs and poultry meat in unit time and unit area.
3) Maximum land utilisation is possible as poultry house can be constructed over the fish pond.
4) The water needed for poultry husbandry practices can be directly obtained from the fish pond.

4.9.2 Fish culture cum pig farming

Inland fish culture in association with pig rearing is a compatible business as it provides an aquaculturist a ready-made source of manure to fertilise the fish farm. The simultaneous farming of fish along with pig is called fish culture cum poultry farming or integrated fish and pig farming.

Principle: The pig manure contains a high amount of nitrogen, phosphorus, ammonium, potassium, etc. In pig excreta N, P, K contents are 0.60%, 0.50%, 0.20% respectively. Thus, it can be used as a substitute for pond fertiliser and supplementary fish feed. As a result, the cost of fish production is greatly reduced.

The main objective of integrated livestock-fish farming is to produce an optimum level of natural protein- rich feed for fish i.e., phytoplankton and zooplankton.

Technique: Fish cum pig farming has now become a popular rural enterprise in developing countries including India. The basic technique involves-

1) **Breed and species selection:** Breeds suitable for integrated pig cum fish farming are large and middle white Yorkshire, Barkshire, Duroc, Hereford, Cross-hybrids of exotic and indigenous pigs and pure local breeds. Large and middle white Yorkshire and their crosses are most preferable breeds of pig for fish cum pig farming system.

   Fish species of high market demand like Indian major carps (rohu, catla and mrigal) and exotic carps (silver carp, grass carp and common carp) are generally preferred for fish cum poultry farming system in India.
2) **Pig house construction:** The pig house must be constructed either at the higher side of the pond nearer to the bank or partially over the

ponds. The pig dung and urine can be either channelized directly into the pond or it can be stored in a pit and applied according to necessity. The height of the pig house should not exceed 1.5 m and it should have a floor space of about 3-4 m^2 per pig. The pig house can be constructed with locally available materials like bamboo sticks and wood.

3) **Stocking density:** The stocking rates vary from 8,000–8,500 fingerlings/ha. The species ratio consists of 40% surface feeders, 30% bottom feeders, 20% column feeder and 10-20 % weed feeders like grass carp. The average dung production is approximately 0.25 tonnes or 500-600 Kg per pig per year. Generally, 30 to 40 pigs are more than enough for fertilising one hectare water body.
4) **Harvesting:** Depending on the demand of the fish in the local markets, partial harvesting of table size fish can be done 6-7 months after stocking. However, final harvesting is done after 12 months of rearing. Fish yield ranges from 5000-6000 Kg/ha/yr.

The pigs attain slaughter maturity (60 - 70 Kg) after six months of rearing, and thus they can be sold out after six months. Around 4,200–4,500 Kg pig meat production can be attained per year in the integrated fish cum pig farming system.

Benefits of fish cum pig farming: The following are the advantages of integrated fish and pig farming-

1) Pig manure is rich in nitrogen, phosphorus, potassium which promotes the growth of natural fish food like phytoplankton and zooplankton. Thus, no additional cost is required for fertilisers and feeds of fish.
2) Fish cum pig integrated farming system results in high production of fish, and pig meat in unit time and unit area.
3) The water needed for pig husbandry can be directly taken from the fish pond.
4) No additional land is required for pig husbandry practices. Thus, it ensures high profit through less investment.

4.9.3 Fish culture cum duck farming

Inland fish culture in association with duck rearing is a compatible business as it provides an aquaculturist a ready-made source of manure to fertilise the fish farm. The simultaneous farming of fish along with duck is called fish culture cum duck farming or integrated fish and duck farming.

Principle: Fish culture and duck rearing are mutually complementary. The fish pond provides natural duck feed like aquatic weeds, insects, and mollusks and water surface for duck raising. On the other hand, duck droppings are rich in nitrogen, phosphorus and potassium. Thus, it provides manure to ponds and hence either directly or indirectly (growth of planktons, the natural fish food) promote fish production. In addition, ducks loosen the pond bottom with their dabbling and help in the release of nutrients from the soil, aerate the water, while swimming and keep water plants in check. Thus, fish-duck integrated farming reduces the costs of fertilisers, formulated feeds for fish farming and maintenance of fish ponds.

Technique: Fish cum duck farming has now become a popular rural enterprise in developing countries including India. The basic technique involves-

1) **Breed and species selection:** Breeds suitable for integrated duck cum fish farming are Indian runner, styles, mete, and megaswari.

   Fish species of high market demand like Indian major carps (rohu, catla and mrigal) and exotic carps (silver carp, grass carp and common carp) are generally preferred for fish cum poultry farming system in India.

2) **Duck husbandry practices:** The following three types of duck farming practice are employed in the integrated fish-duck farming system:
   a) **Raising in open water and pens:** In this method, a large group of ducks (~1000) are allowed to graze in a large water body such as lakes and reservoirs during the daytime, but are kept in pens at night.
   b) **Raising ducks in duck house:** In this method, a duck shed is constructed either near the fish ponds or partially over the fish pond. Duck shed can be constructed using bamboo or any other similar cheap material. Duck droppings are either allowed to fall directly in the fish pond or they are channelized into the pond during the cleaning of duck shed.
   c) **Raising ducks in water:** In this method, the embankments of the fish ponds are partly fenced with a net to form a wet run. This will prevent ducks from escaping the pond, while allows fish to enter into the wet run.

3) **Stocking density:** The stocking rates vary from 8,000–8,500 fingerlings/ha. The species ratio consists of 40% surface feeders, 30% bottom feeders, 20% column feeder and 10-20 % weed feeders like grass carp. Each duck releases around 125-150gm of dropping per day. Thus, around 200–300 ducks adequate enough to manure one hectare fish pond. The average stocking density of duck in duck shed is about 4-6 ducks/m^2 area.

4) **Harvesting:** Depending demand of the fish in the local markets, partial harvesting of table size fish can be done 6-7 months after stocking. However, final harvesting is done after 12 months of rearing. Fish yield ranges from 3500-4000 Kg/ha/yr when 6 species are cultured and 2000-2600 Kg/ha/yr when 3 species cultured.

Eggs can be collected daily in the morning and evening. Around 12,000 eggs and 500 kg of duck meat can be produced per hectare of a pond in each year. However, as the egg laying capacity of duck decreases after 2 years they must be sold after that period.

Benefits of fish cum duck farming: The following are the advantages of integrated fish and duck farming-

1) Duck droppings are rich in nitrogen, phosphorus, potassium which which promotes the growth of natural fish food like phytoplankton and zooplankton. Thus, in this integrated fish cum duck farming system, no additional cost is required for fertilisers and feeds of fish.
2) The water surface of the pond is used for duck raising. Thus, no additional land is required for duckery activities.
3) Ducks feed on predators and allow fingerlings to grow. In addition, ducks help in the release of nutrients from the pond bottom soil with their dabbling, aerate the water, while swimming and keep water plants in check. Thus, fish-duck integrated farming reduces costs of maintenance of fish ponds.
4) Fish cum duck integrated farming system results in high production of fish, duck eggs and duck meat in unit time and unit area.

4.9.4 Paddy cum fish culture

Inland fish culture in association with paddy cultivation is a compatible business as it provides maximum utilisation of land resources or paddy fields. The simultaneous farming of fish in paddy fields is called fish culture cum paddy farming or integrated fish and paddy farming.

Principle: In many areas, the paddy fields remain under water for 3 to 8 months in a year. The idea behind paddy cum fish farming is to use these fields for cultivation of fish. It is advantageous as overall rice production is increased in this method due to soil fertilisation by the fish excreta. Thus, fish-paddy integrated farming is an ideal method for simultaneous production of grain and animal protein on the same piece of land. It has been estimated it is possible to obtain a fish yield of about 2.2 – 2.4 million metric tonnes from the rice fields (Coche, 1967).

Technique: Fish cum paddy farming is an old practice in several countries like Japan, Malaysia, Italy, China and India. The basic technique involves-

1) **Varieties and species selection:** Rice varieties suitable for integrated fish cum paddy farming are IB-1, IB-2, AR-1, 353-146 (Assam), AR 61-25B, PTB-16 (Kerala), Jaladhi-1, Jaladhi-2 (West Bengal) and Thoddabi (Manipur).

   Fish species which tolerant to shallow water, high temperature (up to 350 C), low DO and high turbidity like Indian major carps (rohu, catla and mrigal), minor carps (*Labeo bata*, *Labeo calbasu*, *Puntius japanicus*, *P. ticto*, etc.) and air breathing fishes (*Anabas testudineus*, *Clarias batrachus*, *Channa striatus*, *Channa punctatus*, *Heteropneustes fossilis,* etc.) are generally preferred for fish cum poultry farming system in India.

2) **Site selection:** The site selected for integrated fish cum paddy farming should have a rainfall of about 80 cm. The selected site must have a uniform contour and high water trapping capacity. Good drainage system and flood free area are the other parameters to be considered during site selection.

3) **Stocking:** The fish rearing period varies from 3-6 months, while the paddy rearing period varies from 5-7 months. The average stocking density of fish in fish cum paddy farming is about 2500 fingerlings/ha area.

4) **Method of cultivation:** Following two methods are commonly used for fish culture in rice fields:

   a) **Simultaneous or concurrent culture:** In this method, both paddy and fish are cultivated together. Depending on the type of rice and species of fish to be cultured, it is necessary to maintain a water depth of 5-25 cm. A stocking density of 2000 fingerlings/ha is generally recommended. However, this method has a few limitations like:

- Use insecticides and herbicides for paddy is not possible as these are toxic to fish.
- Cultivation of fish like grass carp, common carp and tilapia are not possible as they often feed on rice seedling or uproot the rice seedlings.
- A high water level cannot be maintained for the fish as it may submerge the paddy.

b) **Rotation culture:** In this method, after the harvesting of rice, fish are cultivated is the same paddy field. After rice harvesting, the field is converted into a temporary fish pond by maintaining a water depth up to 60 cm. As the two activities- fish culture and paddy cultivation are carried out alternatively this method is advantageous over simultaneous method, because:

- It permits the use of insecticides and herbicides for rice production.
- In addition, there is no chance of uprooting of rice seedlings by fish.
- A field can be stocked with a high density of around 6,000 fingerlings/ha.

5) **Management:** The main problem of this method is the use of pesticides as these are toxic to fish. Thus, pesticides like carbomates or furadon must be applied at least 7-15 days prior to fish stocking. It is better to remove the weeds from the paddy fields manually. Predatory and weed fishes can be removed either by netting or by applying Mohua oil cake at a rate of 250 ppm.

Application of manure (5000 kg of cow dung per hectare) and supplementary food consisting of rice bran and groundnut oil cake is also very effective for better fish growth rate in integrated fish cum paddy farming system.

6) **Harvesting:** In this method, fish culture is generally carried out for 3-4 months. During this period, a production rate of 700-1000 kg/ha can be obtained through proper manuring and supplementary feeding. Paddy harvesting is normally made in the last part of September and October. The paddy production rate in this integrated system varies from 59.4-65.4Q/ha (Singh and Gautam 2016).

Benefits of paddy cum fish culture: The following are the advantages of integrated fish and paddy farming-

1) Maximum and economical utilisation of land or paddy field is possible. It involves the production of fish from the paddy field.
2) This method has high rice yield by 5-15%, due to the fertilisation of paddy fields by the fish excreta.
3) Fish-paddy integrated farming is an ideal method for simultaneous production of grain and animal protein on the same piece of land.
4) Unwanted aquatic weeds and filamentous algae compete for the nutrients, and thus often reduce rice yield up to 50%. Cultivation of tilapia and common carp in paddy field control these aquatic weeds.
5) Cultivation of murrels and catfishes control the insect pests of rice like stem borers.
6) Fish also feed on mosquito larvae, snails, etc. Thus, fish cultivation in paddy fields helps in controlling the intermediate hosts of parasitic diseases like malaria, bilharzia, schistosomiasis, filaria and yellow fever.

4.10 CONSTRUCTION OF FISH POND AND ITS MANAGEMENT

4.10.1 Fish pond

The artificial lake or reservoir that is stocked with fish in aquaculture practices for commercial production of fish or for recreational fishing is called a fish pond. Beside, pisciculture a fish pond also fulfils the needs of water for irrigation and livestock.

4.10.2 Types of fish pond

Freshwater fish ponds can be classified on the basis of their purpose, that is, whether it is used for spawning or stocking, materials used in construction and the method of construction (source: http://agropedia.iitk.ac.in/).

A] On the basis of purpose:

1) **Hatching pond:** These are mainly small sized ponds. As the name suggests this type of ponds are mainly used for hatching of fertilised eggs.
2) **Spawning pond:** Like hatching pond, these are also small sized ponds where brood fish are placed for spawning.
3) **Nursery pond:** These are larger pond for newly hatched fry. Stocking is generally done after removal of predatory and weed fish and satisfying the physicochemical nature of the water and plankton growth. Thus, these are seasonal and they generally dry up during summer.
4) **Rearing pond:** In this type of pond juvenile fish or fry are raised till they grow into fingerlings prior to release into stocking the pond. In other words, it is a pond in which the young are reared.
5) **Stocking pond:** These are used for growing fish for 6 to 12 months or till they attain marketable size. The size and shape of the pond depend on the size and shape of the land available. Stocking of fingerlings in stocking pond symbols the beginning of a production cycle.

B] On the basis of purpose:

1) **Earthen ponds or excavated ponds:** This is the most common type of pond. It is constructed entirely from soil materials. These are constructed by removing soil either by manual labour or by dozers from an area up to a required depth. The main advantage of this type of pond is that it is relatively cheaper to construct and fish live in a natural environment.
2) **Walled ponds:** The wall of this type of pond is surrounded by blocks, brick, wooden planking, or concrete.
3) **Lined ponds:** These are similar to earthen ponds the only difference is that the wall of this type of pond is often lined with a plastic, rubber sheet, polyethylene, bitumen, clay, brick, cement-concrete, etc. to make it impermeable. Thus, unlike the unlined pond, the lined ponds have low or even no seepage loss and high storage capacity.

C] On the basis of construction method:

1) **Excavated pond:** These are constructed by removing soil either by manual labour or by dozers from an area up to a required

depth. These types of ponds are usually undrainable and their water source is mainly rainfall, surface runoff or groundwater. They are mostly used in aquaculture.

2) **Embankment pond:** These are constructed simply by building one or more dykes above ground level to hold water.

D] According to the site:

1) **Sunken pond:** The floor of this type of pond is generally below the level of the surrounding land. Thus, it is a completely dig out type of pond. It is not drainable or partially drainable and fed by groundwater, rainfall, surface runoff, pumping, etc.
2) **Barrage pond:** They are created by building a dyke or wall across a small stream. It is drainable through the river bed in which water enters the pond through an inlet from a spring, reservoir, lake, etc. and flows out through the outlet.
3) **Diversion pond:** It is constructed by bringing water into the pond from another source (e.g., spring, lake, reservoir, etc.) through a diversion canal either by gravity or by pumping.

4.10.3 Construction of fish pond

A fish pond is a controlled artificial water reservoir that is stocked with fish aquaculture practices for commercial production of fish or for recreational fishing. The size of a typical earthen fish pond should be 300m^2. One of the major principles behind the successful pisciculture is the proper construction of the fish pond and its proper management. Poorly constructed ponds are hard to manage as water levels may dramatically change if there is leakage or if the watershed area is not large enough, while shallow areas may cause aquatic weeds to grow and spread rapidly. The main physical factors need to be considered during pond construction are the land area, water supply, water outlet and the soil.

1) **Site selection:** Before constructing a fish pond, a proper site selection is the most important step (Kumar 1992). The area where a pond to be constructed must have the following features:
    a) The selected land for pond construction should be relatively level. A sharply sloped land is not suitable for fish pond construction. Generally, a slope of about 1% is ideal.

b) Soil with a large amount of gravel or rocks either on the surface or mixed in is not a suitable site for pond construction.

c) To prevent water leakage, soil that is to be used to build the pond dykes must contain at least 20% clay.

d) The selected site should not be prone to flooding. The selected area should not be subject to pollution due to runoff from the adjacent factory, tea garden, agricultural fields, domestic sewage, etc.

2) **Shape and size:** Square and rectangular shaped ponds are commonly used for fish farming (Fig. 4.9). However, it can have a different shape according to the size and shape of the available land. An area of approximately 300m^2 is ideal for a family pond. Larger ponds must have a surface area of around 1 acre or more for a good pisciculture.

3) **Depth:** Pond depth is an important factor, a stocked fish pond depth should be between 6 and 8 feet. The maximum depth should not exceed 10 to 12 feet. Both low and high depth has disadvantages like a pond with a depth less than 6 feet increases the chances of aquatic vegetation problems, while depths greater than 12 feet are not suitable for good fish production as D.O concentration decreases with depth, low bottom temperature, etc.

4) **Water supply:** The most common sources of water used for aquaculture are surface water (streams, springs, lakes) and groundwater (wells, aquifers). Wells and springs are generally preferred for their consistently high-quality water. Thus, ponds can be fed by groundwater, rainfall, surface runoff or from other water bodies such as a stream, a lake, a reservoir, irrigation canal, etc.

5) **Digging the pond inner area:** After deciding the depth of the pond, digging can be started accordingly. Both manual labour and dozers can be used for digging. After deciding the size and before starting digging, pegs are placed at the inner toes in the four bottom corners. The toe is the point where the inner dyke slope meets the pond bottom. It can be easily calculated by multiplying the desired slope of the dyke and the desired pond depth. For example, at the deep end, the inner toes will be 80 cm x 2 = 160 cm, while at the shallow end the inner toes will be pegged at 75 cm x 2 = 150 cm (source: http://agricoop.nic.in/).

6) **Sealing the pond bottom:** If the dyke or pond bottom soil is highly permeable that is sandy type, a core trench must dig under the dykes surrounding the pond. The packing of this core trench then can be done with impermeable clay which will prevent water loss through seepage across pond sides or the bottom. The most commonly used pond sealant is bentonite clay.

The pond bottom should also have a slope so that water varies in depth along its length as different fish species prefer different water depth. A slope of around 2-5% is ideal for the pond bottom. In addition to this, after digging the pond bottom must be smooth out as it will help in netting operation during harvesting.

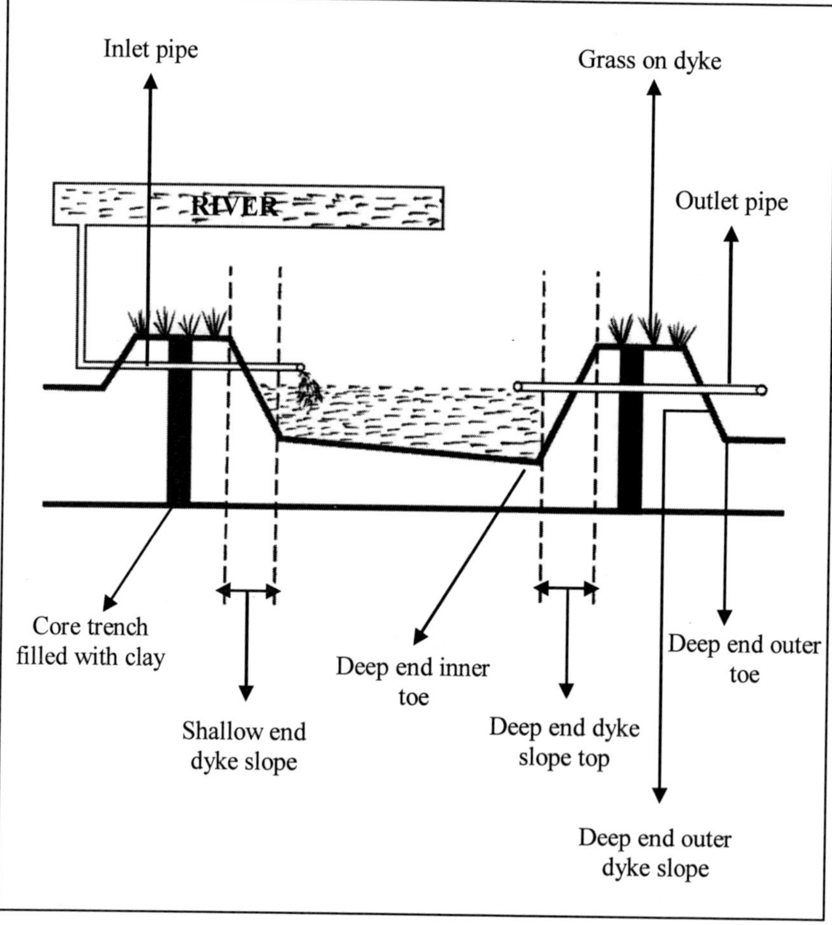

Fig. 4.9: Cross section of earthen pond showing bottom slope, dyke slope, inner and outer toe.

4) **Construction of dykes:** The most important component of a pond is its walls called dykes. The soil excavated from the pond area through digging is generally used to construct the dykes. Topsoil (sandy and rocky) and soil, which contain roots, grass, sticks or leaves are not good material for dyke construction, so these should not be used for dyke construction.

   The pond dyke should be 30cm above the water level. If catfishes will be farmed, the dyke should be 50cm above the water level due to their jumping behaviour. The dyke must have an inner and outer slope. The best slope for pond dyke is one that rises 1 metre in height for every 2 metre in length. After construction of dykes, grasses can be planted on dyke top to protect soil erosion from the dyke during raining or flood.

5) **Installing of drainage system:** A water outlet channel must be installed after the dyke has been constructed. First, a channel is cut across the dyke just above the optimum water level. The pipe is then fitted in the ditch at a proper slope through the dyke (not less than 1%) so that water can easily flow out of the pond when it reaches above the desired level. The outlet pipe must be equipped with a fine nylon screen to prevent the escape of fish from the pond.

   Many ponds also have an inlet pipe to carry water by pumping from the river or other water reservoirs during the dry season. Like outlet pipe, the inlet pipe is also equipped with a fine nylon screen to prevent unwanted entry of weed and predatory fish into the pond.

4.10.4 Management of fish pond

Ponds will provide excellent fishing opportunities if they are properly managed. The keys behind the scientific and successful fish farming are proper planning, construction, and management. Pond management includes fish stocking, pond design, pond assessments, weed control, clearing muddy ponds, etc. or more correctly everything which promotes fish production. Some of the common management strategies for better fish production in ponds include the following:

1) **Removal of unwanted fish:** Fish populations often become imbalanced or contaminated with unwanted species like predatory fish and weed fish. The fish species (e.g., *Wallago attu*, *Nandus nandus*, *Channa striatus*, *Clarias batraches*, *Heteropneustes*

fossailis, etc.) that feed on the spawn, fry and fingerling of cultured fishes are called **predatory fishes**. The unwanted uneconomical, small sized, naturally occurring fish (e.g., *Puntius sp.*, *Amblypharyngodon mola. Colisa sp.*, *Rasbora sp.*, etc.) found in a pond or other water bodies are called **weed fishes**. They get into stocking ponds through inlet water or during stocking. These have many disadvantages like

- They breed easily in confined water prior to the onset of carp breeding. Thus, the size of predatory fishes will be bigger than the size of the carps at a particular time and they directly prey on them.
- These unwanted fish compete for food, space and DO. This results in poor growth of desired fish.
- Some predatory fish have the habit of burrowing in the mud bottom.
- Their complete eradication using physical methods is very difficult.
- Weed fishes have high fecundity and they can even breed in the absence of rains in captivity.

Their eradication prior to the stocking of carps or desired fish is necessary. They can be eradicated by

a) **Repeated netting:** Repeated netting and catching of predatory fish can be done for their control. But, it is not possible to catch all predatory fishes simply by netting.

b) **Dewatering:** During off-season, dewatering followed by sun drying is the most effective method to control both predatory and weed fish.

c) **Use of fish toxicants:** Fish toxicants can be used prior to fish stocking to kill all predatory and weed fish. Some of the common examples of fish toxicants are:

   - Chlorinated hydrocarbons: Aldrin, endrin, Dieldrin
   - Organophosphates, thiometon, DDVP and phosphamidon
   - Plant-derived: Mahua oil cake, tea seed cake. The mahua oil cake is the most commonly used fish toxicant in India. It kills fish at 200-250 ppm in 6-10 hours. Its main advantage is that its toxicity lasts for only 15 -20 days in water and after this period, it subsequently converted into pond manure.

2) **Harvesting:** A balanced pond fishery can be established with the initial stocking according to the area. Proper management of harvesting is necessary to maintain that balance in the pond. Harvesting can be drain-and-seine (total) or the multiple (partial) harvest method. Harvesting can be started when the fish reaches the marketable size and in general, they will be big enough to harvest in about six months.
3) **Artificial feeding:** All ponds produce some natural food (zooplankton and zooplankton) for fish, but sometimes it is not adequate. Formulated fish feeds, which are rich in protein, carbohydrates, fats, vitamin and minerals (macro and micro) in pellet form, are very common.
4) **Fertilising the pond:** Water fertility determines the overall productivity of the pond. Fertiliser increases pond productivity by stimulating the growth of planktons. Fertiliser commonly used includes animal manures, compost or chemical fertilisers (Urea, calcium superphosphate, ammonium sulphate, etc.). If the soil is acidic, the addition of lime or wood ashes is also recommended. Generally, 10-20Kg of lime or 20.40Kg of ashes is used for each 100m^2 of pond bottom.
5) **Record Keeping:** Accurate records keeping of numbers and sizes of fish stocked and harvested in the pond along with physicochemical parameters of the pond will help in evaluating the status of the production. This helps in taking an additional management if needed.
6) **Eutrophication:** It is the depletion of oxygen in a water body, which kills aquatic animals. It occurs mainly due to excess nutrient enrichment, which in turn induces explosive growth of plants and algae, decaying of which consumes oxygen from the water. In general, fish kills can be prevented by properly controlling nutrient inputs and overabundant aquatic vegetation.
7) **Aquatic weed control:** Weeds generally do not have adverse effects if the vegetation covers less than 20-25% of the pond's surface. But when it becomes overabundant, covering more than 20-25% of the pond they must be controlled by
    a) Physical (hand pulling, cutting) or
    b) Biological (cultivating grass carp) or
    c) Chemical (herbicides) means.

8) **Managing other animals:** Beavers, muskrats, crayfish, snakes, and turtles all can affect your ponds in their own way. The use of traps in the most effective, practical and environmentally safe method for control.
9) **Enhancement techniques:** In addition to managing, several other techniques like a) adding fish shelters/habitat, b) supplemental feeding, c) checking and adjusting water levels, and d) aeration can improve pond fish farming.

4.10.5 Fertilisation and manuring of fish pond

Application of nutrients in the ponds for promoting the growth of phytoplankton and rate photosynthesis is called fertilisation. Since zooplankton and other aquatic animals including fish fed on phytoplankton, fertilisation increases the abundance of natural fish food in a pond. Fertilisers can be broadly classified into two types- inorganic and organic fertiliser.

A] Inorganic fertilisers: Inorganic fertilisers mean chemical fertilisers. According to the composition, the chemical fertilisers can be divided into three groups, namely, nitrogenous, phosphoric, potassic and complex fertilisers. The advantages of inorganic fertilisers are its exact constituents and fast effect, no consumption of DO, small application amount and convenient operation (FAO, 1985).

Types of inorganic fertilisers: Pond fertilisers are available in liquid, granular, or powdered forms. Fertilisers can be

- Nitrogenous fertilisers (Liquid ammonia, ammonium sulphate, calcium ammonium nitrate, ammonium sulphate nitrate, urea, etc.),
- Phosphoric fertilisers (Calcium superphosphate, triple super phosphate, etc.) and
- Potassic fertiliser (Potassium chloride, potassium sulphate, etc.).

Time of fertilization: Fertilization can be started in the spring when water temperature is stabilized at 60^0F. In addition to this, the simplest method of knowing when to fertilise the pond is to examine the clarity of the water. Light penetration can be measured using a Secchi disk. If the sunlight penetrates the water 18 inches or more, a fertilization program can be started and continued all summer.

Importance of liming before fertilization: Acidic soils will bind or hold the phosphorus contained in the fertiliser and the phosphorus will not be dissolved into the water column. Thus, if the soil is acidic, it is very important that liming of pond water should be done before application of fertiliser.

Rate of fertiliser application: Since physicochemical parameters of both soil and water vary from pond to pond, the type of soil in the pond will determine the amount of fertiliser to be applied.

1) **Liquid fertiliser (10-34-0 or 13-37-0):** The number denotes the concentration of nitrogen-phosphorus-potassium (NPK) in the fertiliser. One gallon per surface acre is generally recommended. If after 7 days no bloom develops, one quart is applied for every 7 days until bloom develops.
2) **Granular fertiliser (0-20-0):** Granular fertiliser like triple super phosphate is the most economical one. Generally, 24 pounds is recommended for one surface acre. If after 7 days no bloom develops, 10 pounds is applied for every 7 days until bloom develops.
3) **Water soluble fish pond fertiliser (10-52-4):** Application of 4 pounds per acre can increase the number of fish in your pond by three to four times.

Methods of the application of fertiliser: Spreading over the entire pond surface homogeneously is preferable for liquid fertiliser. As per the formulations prescribed by the manufacturer some liquid fertilisers must be diluted prior to application, while others can be directly applied into the fish pond. Since, most of the liquid fertilisers are heavier than water, and thus if they are applied at one spot in the pond, the nutrients will sink to the bottom and become bound in mud and sediment. Alternatively, the container of liquid fertiliser can be kept under water with an open lid. Liquid ammonia will then diffuse out slowly with the course of time (Brunson et al. 1999). On the other hand, in the case of granular fertiliser, it is important to prevent the direct contact of granules with the mud. A significant amount of the phosphorus will be trapped in the mud and it will be completely wasted.

B] Organic fertiliser/ manure: Organic manures mainly refer to the excrements of farm animals. Organic manures may be converted to nutrients by microbial decomposition, which in turn promotes phytoplankton growth. Organic manure increases overall productivity in three different ways-

1) The organic manure stimulates the propagation of bacteria in a rapid scale. The bacteria then decompose the organic materials, mineralizing them into nutritional inorganic materials, which in turn can be utilised by phytoplankton (producers).
2) Organic fertiliser like animal manure provides nutrients and attachment sites for bacteria and other microscopic organism.
3) Organic manure like green manure and undigested food in animal manure are digestible, and thus they provide direct nutrition when eaten by fish.

Types of organic manures: The common nature organic manures used in pisciculture are- 1) Faeces and urine of livestock (cattle manure, pig and poultry manure) which contains organic matter and other nutritional elements like nitrogen, phosphorus and potassium, 2) Nightsoil (human excreta and urine), 3) Silkworm dregs, 4) Green manures (decomposed plants or grass) and 5) Compost (green manure + animal excreta).

Methods of the application of organic manures: Types of organic manure determine the method of organic manure application.

- **Livestock manure:** Application procedure of livestock manure and green manure is almost similar (FAO 1985). Before application, the manures are first heap at a corner of the pond and then applied in small heaps in shallow water during sunny days. This will help rapid manure decomposition and spreading in a water body.
- **Nightsoil:** In the case of nightsoil application, it must be diluted with water (1:2). The dilution is then sprayed along the pond dykes once a day, the amount of which depends on the fertility of water.
- **Compost:** In the process of compost application, after fermentation, the compost is taken out and given a flush. The liquid is collected and the residues are removed away. The liquid is then evenly sprayed into ponds. For small ponds, the liquid can be spread around the dykes. On the other hand, in the case of big ponds, the farmer can spray the liquid evenly in ponds using a boat. The main advantage of compost application is that the nutrients in the compost can be quickly absorbed by phytoplankton.

One of the major advantages of compost application is that it consumes less dissolved oxygen (DO) since the organic materials are already decomposed during fermentation.

4.11 REFERENCES

Alikunhi KH (1957) Fish culture in India. Fm. Bull. Indian Coun. Agric. Res. 20: 144.

Alikunhi KH, Sukumaran KK, Parameswaran S (1971) Studies on composite fish culture: Production by compatible combinations of Indian and Chinese carps. Journal of Inland Fisheries Association. 1(1): 26–57.

Bhatnagar A and Devi P (2013) Water quality guidelines for the management of pond fish culture. International Journal of Environmental Sciences. 3(6): 1980-2009.

Brunson MW, Stone N, Hargreaves J (1999) Fertilization of Fish Ponds. SRAC Publication No. 471.

Chakrabarty RD, Singh SB, Rao NGS (1979) The evolution of the technique of composite fish culture at CIFRI. In: Souvenir of CIFRI, Golden Jubilee Year of ICAR.

Coche AG (1967) Fish culture in rice fields: a worldwide synthesis. Hydrobiologia. 30(1): 1–44.

FAO (1985) Training manual integrated fish farming in China. Regional Lead Centre in China Asian-Pacific Regional Research and Training Centre for Integrated Fish Farming Wuxi, China.

Ghosh A, Saha SK, Roy AK, Chakraborti PK (1985) Carp production using domestic sewage. Aquacult. Ext. Man. Inland Fish. Res. Inst., Barrackpore, (8):19p.

Gupta MV and Noble F (2001) Integrated chicken–fish farming. Halwart M,. Gonsalves J, Prein M, eds. Integrated agriculture – aquaculture: A primer, FAO Fisheries Technical Paper No. 407, FAO, Rome, Italy.

Izquierdo MS, Fernandez-Palacios H, Tacon AGJ (2001) Effect of broodstock nutrition on reproductive performance of fish. Aquaculture 197(1-4): 25-42.

Kumar D (1992) Fish culture in undrainable ponds. A manual for extension. FAO Fisheries Technical Paper No. 325, FAO, Rome, Italy, 239 p.

Kumar V and Karnatak G (2014) Engineering consideration for cage aquaculture. IOSR Journal of Engineering. 4(6): 11-18.

Kureel S, Prakash S, Sharma A, Ambulkar R (2013) Marketing of ornamental fishes in national capital region, India. Progressive Agriculture. 13(2): 73-77.

Ngueku BB (2014) The design and construction of Aquaria. International Journal of Fisheries and Aquatic Studies. 2(3): 01-04.

Novotny AJ (1975) Net-pen culture of Pacific salmon in marine waters. Marine Fisheries Review. 37(1): 36–47.

Odd-Ivar Lekang (2013) Aquaculture Engineering, 2nd edn. John Wiley & Sons, Inc, Hoboken, New Jersey, pp. 130-165.

Schartl M (2014) Beyond the zebrafish: diverse fish species for modeling human disease. Disease models & mechanisms. 7(2): 181–192.

Singh JB and Gautam US (2016) Paddy cum fish culture: beneficial technology in low land areas of rice. Agriculture. 6(2): 84-86.

Sukumaran KK (1976) Progress report of the Karnal sub-centre of Haryana for the period November, 1972 to August, 1975, Proc. third workshop on the all India coordinated research project on composite fish culture and fish seed production, 26-27, February, 1976, CIFRI, Puri.

Torrissen O, Olsen RE, Toresen R, Hemre GI, Tacon AGJ, Asche F, Hardy RW, Lall S (2011) Atlantic Salmon (*Salmo salar*): The 'Super-Chicken' Of The Sea?. Reviews in Fisheries Science. 19 (3): 257–278.

Weimin M (2004a) Cultured aquatic species information programme *Hypophthalmichthys nobilis*. In: FAO Fisheries and Aquaculture Department, Rome.

Weimin M (2004b) FAO cultured aquatic species information programme *Ctenopharyngodon idellus* (Valenciennes, 1844).

*** ***** ***

Chapter 5

FISH DISEASES, FISH SPOILAGE AND FISH PRESERVATION

Rupak Sharma[1], Baby Singha[2,*]

[1]Department of Zoology, Aryan Junior College, Silchar, Assam, India
[2,*]Department of Zoology, G. C. College, Silchar, Assam, India

Chapter highlights:
Infectious diseases pose one of the most significant threats to successful aquaculture. Fish ailments can be due to viral, bacterial, fungal and protozoan infections. In addition to this, adverse environmental condition, nutritional disorders, and genetic defects are also responsible for noninfectious diseases like gas bubble disease, swim bladder stress syndrome, sunburn disease, constipation, physical injury, etc. Post-harvest loss (PHL) is a major concern in fishery industry throughout the world. Post-harvest fish spoilage occurs mainly due to enzymatic, microbial and chemical action. However, different post-harvest processing and preservation techniques like drying, freezing, salting, smoking, canning, etc. help in lengthening shelf life while maintaining their flavour, taste, and nutritive value. The fish represents a valuable source of proteins and nutrients and thus, the most common use of fisheries resources is food. In addition to food, the fish industry also yields a number of byproducts like body oil, liver oil, fish emulsion, fish glue, isinglass, fancy articles and many more for commercial purposes. Besides this, several compounds extracted from fish are also employed as component in many official medicines like tetrodotoxin (TTX) isolated from puffer fish is as an extraordinary narcotic and analgesic.

5.1 FISH DISEASES

Disease is a prime agent affecting fish mortality, especially when the fish are young. There are literally hundreds of afflictions that can affect the fish health, and thus infectious diseases pose one of the most significant threats to successful aquaculture.

Fish ailments can be broadly categorised into five general categories including a) viral infection, b) bacterial infections, c) fungal infections, d) parasitic or protozoan infections, and e) physical ailments and wounds (Fig. 5.1).

5.1.1 Viral infection

Viruses are small infectious agents that are capable of replicating themselves inside the living cells. Over 125 different viruses have been documented in fishes, but most of them have been in aquacultured food fishes (Noga 2010). Some of the common viral diseases are infectious pancreatic necrosis (IPN), viral haemorrhagic septicaemia (VHS), infectious haemopoietic necrosis (IHN), spring viraemia of carp (SVC), pike fry virus disease (PFVD), channel catfish virus (CCV) disease, lymphocystis disease (LD), papillomatosis (cauliflower disease of eels), carp pox etc (Table 5.1).

5.1.2 Bacterial infection

Fish are more susceptible to bacterial pathogens, particularly when they are physiologically unbalanced, nutritionally deficient, or there are other stressors, like poor water quality and overstocking. Most of the bacterial diseases are generally characterised by red streaks or spots and/or swelling of the abdomen or eye. In addition to good management practices, antibiotics such as penicillin, amoxicillin, or erythromycin are commonly used for treating bacterial infection in fish. Some of the most common bacterial diseases are dropsy, fish tuberculosis, fish vibriosis, tail rot & fin rot, columnaris or cotton-mouth, etc.

5.1.3 Fungal infection

As fungal spores are naturally found in all fish aquariums and water body, they can quickly colonise and cause diseases in stressed or injured fish.

Thus, poor water quality, poor hygiene, fish that are injured, old, or have other diseases can intensify the situation, and thus lead to an increase in fungal infections. Common fungal infections often look like grey or white fluffy patches. Branchiomycosis, icthyophonus, saprolegniasis or cotton wool disease, exophialasis, cerebral mycetoma are the most common fungal infections in fishes.

5.1.4 Other parasitic infection

Parasites can be internal or external. The majority of the diseases causing fish parasites include protozoal parasites. In addition to this, myxozoan (a group of aquatic parasitic cnidarian animals), monogenetic and digenetic trematodes, cestodes, parasites belonging to phylum Acanthocephala, leeches and copepods, etc. are the others which routinely infect fish. The common parasitic diseases can be treated most effectively by formalin or malachite green, or a combination of the two, copper, methylene blue and quinine hydrochloride in the right dosage. Most treatments generally have copper as an ingredient. Heat treatment around 86°F (30°C) is also highly effective in some cases.

5.1.5 Non-infectious illness

Surrounding environment is mainly responsible for various physical ailments like fish gasping, not eating, and jumping out of the tank results from poor quality water, quality, while nipped fins and bite wounds mainly results from tank mate problems. Some of the common noninfectious illnesses found in fish are as follows:

1) **Physical injuries:** Common physical injuries in fish can be treated with 2% mercurochrome. Additionally, bathing in slightly acidic water also speeds up the recovery if fish tolerate lower pH values. A pH around 6.6 is ideal for healing almost all types of minor physical injuries.
2) **Constipation:** It is a common problem arises in aquarium fish mainly due to unsuitable diet (fibreless diet). Constipation in fish is typically characterised by loss of appetite, swelling of the body bloating and the production of stringy faeces which hang from the fish. In severe cases, constipation often makes the swimming

difficulty in fish. Fish should be fed a varied diet containing a lot of green, fresh, live or frozen foods. Providing dried food that has been soaked in paraffin oil or glycerol or castor oil and live foods or change in diet on a regular basis often minimises the problem.

3) **Tumours:** Most tumours are genetical though tumours can be caused by a virus or bacteria also. Too much hybridisation is the main cause of genetic tumours. Practically all tumours are un-treatable, and thus it is better to remove the fish if it is in distress.

4) **Gas bubble disease:** Supersaturation of dissolved gases like nitrogen or oxygen often causes gas bubble disease in fish. Leaks in a pump or valve systems in hatcheries; dense algal bloom that presumably caused oxygen depletion at night and supersaturation during the day are the common reasons behind this problem.

   Common symptoms generally observed in affected fish are bubbles in the abdominal cavity, eyes, skin, gills, fins, mouth, swim bladder and within the digestive tract. The problem can be solved by good husbandry practices like regular monitoring of DO, avoid algal bloom formation, proper aeration and sufficient water exchange, etc.

5) **Swim bladder stress syndrome (SBSS):** Any type of malfunction of the swim bladder leads to swim bladder stress syndrome. High ambient temperature, high ambient illumination, dense algal bloom and super-saturation during the day are the main causes of SBSS. The appearance of large bubbles of gas in the anterodorsal region and outside the swim bladder is the diagnostic characteristic of swim bladder stress syndrome. Preventive measures are same as those used for the gas bubble disease.

6) **Sunburn disease:** It main occurs in fish which are stocked in shallow, uncovered raceways under intense sunlight. Affected fish develop grey focal circular ulcerative lesions on top of the head, pectoral, dorsal and upper tail fins. The simple way to minimise the problem is the arrangement of sunshades over ponds.

7) **Asphyxiation/hypoxia:** As expected very low level of dissolved oxygen in the water body leads to hypoxia in fish. Fish in hypoxia condition, swim at the water surface, show rapid opercular movement and generally gather at the water inlets and outlets. Common management practices are regular monitoring of DO, management of phytoplankton blooms and reduction of nitrogen in the system.

8) **Alkalosis and acidosis:** Too drop or rise in pH from the tolerable limit also causes abnormalities in fish. In alkalosis, affected fish show corroded skin and gills, while in acidosis affected fish show rapid swimming movements, gasping and increased mucus secretion. Lime application and flushing pond bottom before stocking are the common preventive measures employed in acidosis.

9) **Ammonia and nitrite toxicity:** Ammonia along with nitrite is sometimes referred to as one of the "invisible assassins" in aquaria. Common clinical signs are dyspnea, telangiectasia, gill damage, excessive mucous production, and lethargy. It also has a number of harmful side effects like increased disease susceptibility, suffocation, organ failure.

   The addition of salt at the rate of 1200-1800 kg/ha reduces toxicity. A biological filter may be used to treat water for converting ammonia to nitrate. This nitrate can be then converted to harmless nitrate through nitrification process.

10) **Salinity:** High salinity is also responsible for different types of noninfectious illnesses like progressive emaciation, scale loss, and opaque eye lenses.

11) **Hypercapnia or hypercarbia:** An extremely elevated concentration of CO2 in water is known as hypercapnia. Common clinical signs associated with hypercapnia include narcosis, lethargy or depression, slowed respirations, etc.

12) **Hypothermia and hyperthermia:** Common clinical signs associated with extreme temperature fluctuations are hyperactive, tachypnea, rapid swimming, ruptured blood vessels, listless, lethargy, clamped fins and non-progressive swimming.

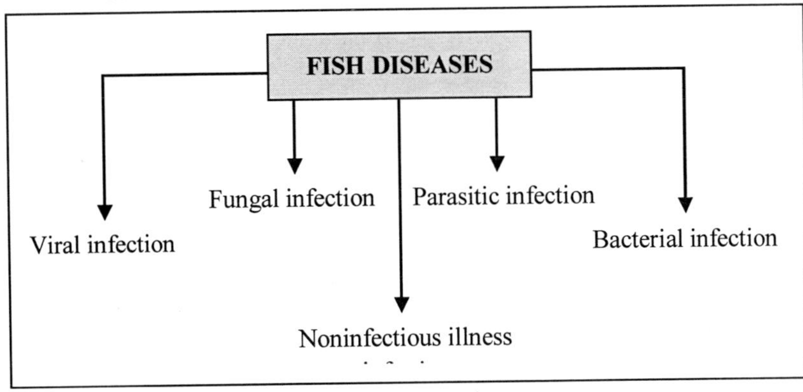

Fig. 5.1: Causes of fish diseases

Table 5.1: Common examples of viral, bacterial, protozoan and other parasitic infection, their signs and symptoms and prevention and control measures.

No	Name of the disease	Causative organism	Signs and symptoms	Prevention and treatment
VIRAL DISEASES				
1	Infectious pancreatic necrosis (IPN)	Infectious pancreatic necrosis virus (Family-Birnaviridae), a dsRNA virus (Wolf et al. 1960). Mainly affects young salmonids, such as trout or salmon.	1. Common signs and symptoms include abdominal swelling, anorexia, showing a spiral type of swimming, darkening of the skin, and trailing of the faeces from the vent. 2. On necropsy, internal damage (viral necrosis) to the pancreas and thick mucus in the intestines are often found.	1. Currently, no method is available to remove the virus from infected fish. Good husbandry measures, such as high water quality, low stocking density, the introduction of virus free fingerlings, avoidance of using nets and other equipment without disinfection helps in reducing the disease incidence. 2. Eradication of virus from infected fish farms could be possible through the slaughtering of all infected fish followed by disinfection of water bodies.

2	Viral haemorrhagic septicaemia (VHS) or egtved disease	Viral Haemorrhagic Septicaemia Virus (Family- Rhabdoviridae), a ssRNA virus (Jenson 1963). Infects both fresh and saltwater fish, predominantly rainbow trout.	1. Common signs and symptoms include haemorrhaging of internal organs, skin, and muscle. 2. Appearance of bulging eyes, bloated abdomens, bruised-looking reddish tints to the eyes, skin, gills and fins.	Same as IPN
3	Infectious haemopoietic necrosis (IHN)	Infectious hematopoietic necrosis virus (Family- Rhabdoviridae), a ssRNA virus. Mainly affects young salmonids, such as trout or salmon.	1. Clinical signs include abdominal distension, bulging of the eyes, skin darkening, abnormal behaviour, anaemia, fading of the gills, haemorrhage in several areas like the mouth, behind the head, the pectoral fins and muscles near the anus. 2. Diseased fish weaken eventually floating "belly-up" on the surface of the	Same as IPN

			water. Necrosis is common in the kidney and spleen, and sometimes in the liver.	
4	Spring viraemia of carp (SVC) or swim bladder inflammation	*Rhabdovirus carpio* (Family- Rhabdoviridae), a ssRNA virus. Infect a wide variety of fish species, including silver carp, grass carp, crucian carp, bighead carp, ornamental koi, and common carp.	1. Clinical symptoms include external haemorrhaging, pale gills, and ascites. Affected fish often appears pale having dark skin. 2. Haemorrhage and inflammation of the swim bladder which leads to abdominal distension, lethargy, imbalance, swimming on their side and sinking.	1. Along with the control measures for IPN, elevation of water temperature above 19–20^{0}C can stop or prevent SVC outbreaks. 2. A safe and effective vaccine is not currently available, but experimentally inactivated preparations and live attenuated vaccines, chemotherapy using methisoprinol (inhibits the replication of SVCV), culturing of resistant breed (Krasnodar strain of common carp) have given encouraging results.
5	Channel catfish virus (CCV) disease	Ictalurid herpesvirus 1 (Family-	1. Infected fish shows haemorrhages on the fins	1. Common control measures include avoidance of high

		Alloherpesviridae). Commonly infect catfish, particularly the channel catfish (*Ictalurus punctatus*).	and musculature. 2. Affected fish tend to swim erratically or close to the surface, eventually sinking to the bottom.	stocking densities, maintenance of appropriate quarantine and hygiene measures. 2. The virus is sensitive to acidic pH, heat and UV light and is inactivated by pond mud and sea water.
6	Lymphocystis disease (LD)	Genus *Lymphocystivirus* (Family- Iridoviridae). It is a common viral disease of many freshwater and saltwater fish.	1. Development of small white pin-pricks like growths on their fins or skin. 2. The lesions often clump together to form a cauliflower-like growth on the skin, mouth, fins, and sometimes in the gills.	1. Application of human antiviral "Acyclovir" at the rate of 200 mg per 10 US gallons for 2 days gives encouraging results. 2. Some prefer surgical removal of the affected area followed by an antibiotic bath treatment.
7	Papillomatosis (Cauliflower disease of eels)	Viruses of Orthomyoxovirus group. Commonly infect European eels. The disease often affects the	1. The disease is characterised by the development of nodules which is approximately the size of a lentil grain in the snout	No effective treatment is known. Though the disease is not fatal; however, fish with tumours must be removed from the tank to avoid horizontal transmission.

		genus *Anabas*, *Colisa* and *Trichopsis* members of the Anabantidae family.	region. 2. The colour varies from pink to deep red, grey or blackish brown, depending on the degree of blood supply and pigmentation.	
8	Carp pox	Herpes virus namely Cyprinid Herpesvirus 1 (Family- Herpesviridae). It is a common skin disease of cyprinids (carps and minnows).	1. The disease is characterised by the development of semi-translucent lumps which is similar to the blobs of white to pink candle wax. 2. The lumps are raised 1-3mm above the skin on the head, fins and other body surfaces.	1. Just like others, the only effective method of control is by preventing the introduction of infected fish in the tank. 2. Raising the temperature of the water to above 15°C, liming the ponds, or transferring the fish to tank supplied with clear, oxygenated water often accelerated the recovery of fish.
BACTERIAL INFECTION				
1	Dropsy	The disease is caused by gram negative bacteria commonly present in	1. In dropsy, fluid is buildup inside the body cavity or tissues of a fish.	1. Quarantine infected fish to prevent spreading, disinfection with 1 ppm potassium

		aquarium habitats.	2. Common symptoms observed in most cases include swollen belly, bulging eyes, gills become pale, anus becomes red and swollen, ulcers formation in the body along the lateral line, the fish becomes lethargic and stops eating.	permanganate solution or dip in 5 ppm copper sulphate for 2 mins (Gopalakrishnan, 1963), use of antibiotic like chloromycetin or tetracycline in food. 2. The other common preventive measures include regular water changes, Keeping the tank clean, Cleaning the filter regularly, increasing the water temperature, avoiding overcrowding and overfeeding, etc.
2	Fish tuberculosis, piscine tuberculosis, acid-fast disease, granuloma disease	It is caused by the bacterium *Mycobacterium piscium* or *M. marinum*	1. The main symptoms of fish tuberculosis are the loss of scales, loss of colour, lesions on the body, hollow bellied, loss of appetite, and skeletal deformities such as curved spines. 2. Sometimes yellowish or	Along with common preventive measures as mentioned above for dropsy, the most effective treatment known for this disease is to treat with Kanamycin and Vitamin B-6 for 30 days.

			darker nodules may appear in the eyes or body.	
3	Fish vibriosis	*Vibrio anguillarum, V. ordalii, V. damsela,* and *V. salmonicida* are the common species responsible for fish vibriosis. Compare to tropical fish, it is more common in marine animals or brackish water fish.	1. Common symptoms include blood streaks under the skin surface, red spots on the ventral and lateral areas, swollen dark lesions which release bloody pus. 2. Infection may also cause eye problems with cloudy eye, which often lead to pop-eye and eye loss.	Along with common preventive measures as mentioned above for dropsy, the most effective antibiotic treatments known for this disease are kanamycin and chloramphenicol or furazolidone.
4	Tail rot & fin rot	Fin rot is commonly caused by a bacterial species of the genus *Aeromonas, Pseudomonas.*	1. Disintegrating fins or ragged, frayed fins, blood on the edges of the fins, reddened areas at the base of fins. 2. Appearance of skin ulcers with grey or red margins, cloudy eyes.	Along with common preventive measures as mentioned above for dropsy, a good antibiotic is chloromycetin or tetracycline. Bathing the infected fish in 1:2000 solution of copper sulphate for 1-2 minutes is also very effective (Hora and Pillay, 1962)

5	Columnaris or mouth fungus or cotton-mouth	It is caused by the gram-negative, rod-shaped bacterium *Flavobacterium columnare.* It infects freshwater aquarium fish, particularly live bearing fish and catfish.	1. Common symptoms are frayed and ragged fins, ulcerations on the skin, white or cloudy, fungus-like patches, particularly on the gill filaments. 2. Accumulation of mucus in gill and gill becomes light or dark brown in colour.	1. Good antibiotics for treating columnaris are furan-2 and kanamycin. 2. Other preventive measures are similar to dropsy except instead of raising, lowering the temperature of the aquarium to 75°F (24°C) is more effective as columnaris is much more virulent at a higher temperature. 3. Salt treatments like potassium permanganate, copper sulphate, and hydrogen peroxide can also be applied externally to adult fish and fry at low concentration.
6	Mouth rot	It is caused by *Flexibacter columnaris*	1. Initial symptoms include development of characteristic small white to grey marks on the head, fins, or gills.	1. A treatment, bath containing phenoxyethanol is normally curative. Salt treatments like potassium permanganate, copper sulphate, and hydrogen

			2. The lesions that grow generally resemble white, fluffy, fungus-like tufts; however, lesions are coarser and greyer in colour than most fungal infections.	peroxide can also be applied externally to adult fish and fry at low concentration. 2. The most effective antibiotic is oxytetracycline.
FUNGAL DISEASES				
1	Branchiomycosis or gill rot	The disease is caused by *Branchiomyces sanguinis* and *Branchiomyces demigrans.*	1. It is characterised by necrosis of the gills and anorexia. Gill appears eroded and pale. 2. The infected fish exhibits respiratory symptoms and loss of equilibrium. Infected fish show signs of asphyxia, such as gasping. Fish often gathered in groups at water inlet and die.	1. Common preventive measures include reducing overcrowding, levels of ammonia, algal blooms, levels of organic material, water temperature and improving hygiene. 2. Treating infected fishes or pond with malachite green, formalin baths, copper sulphate and benzalkonium chloride dips, and oral methylene blue help in spreading of fungal infection.

2	Icthyophonus disease	The disease is caused by fungus *Ichthyophonus hoferi.* Both freshwater and marine fish suffer from this disease.	1. Common symptoms are bulging eyes, loss of appetite, body cavity exudation, dropsy, protruding scales, weight loss, movement disorder and imbalance, loss of body colours. 2. Development of white nodules in the internal organs like liver, kidneys, spleen, etc. and small nodules on the skin.	1. This is an incurable disease. Common preventive measures are similar to those commonly employed for branchiomycosis. 2. Addition of 1% phenoxethol solution or chloromycetin is also effective.
3	Saprolegniasis or cotton wool disease	Most commonly caused by the *Saprolegnia* species called "water moulds. It is a fungal disease of fish and fish eggs of both fresh and brackish water.	1. Development of cotton-like fungal growth of white to shades of grey and brown colour on the skin, fins, gills, or eyes of fish or on fish eggs are the common symptoms. 2. New lesions are generally white, but over time will	1. A good management practice like good water quality and circulation, avoidance of crowding to minimise injury, and good nutrition is the key method for controlling saprolegniasis. 2. Potassium permanganate, formalin, and povidone-iodine

			become red, brown, or green	solutions are commonly used for treatments. Bath treatment, including NaOH (10-25g/lit for 10-20min), $KmNO_4$ (1g in 100lit of water for 30-90 min), $CuSO_4$ (5-10g in 100 lit water for 10- 30min) are also very effective.
4	Exophialasis	*Exophiala salmonis* and *E. psychrophila* are the two common species associated with exophialasis	1. Fish become darker and lethargic. Infected fish often shows erratic and abnormal swimming behaviour. 2. Development of round yellow to white granulomas in visceral organs like liver, kidney and spleen. Posterior kidney often shows prominent enlargement.	Same as cotton wool disease.
5	Cerebral mycetoma	Cerebral infection by *Exophiala salmonis* in	1. Formation of granuloma with numerous giant cells	Same as exophialasis.

		cutthroat trout (*O. clarkii*) fingerlings. It can cause mortality up to 40% in Atlantic salmon hatcheries.	in the brain and cranial tissues. 2. The infection may extend peripherally to surrounding cranial structures, such as eyes and gills.	
OTHER PARASITIC INFECTION				
1	Fish louse (Argulus)	The disease is caused by a flattened mite-like crustacean *Argulus foliaceus.*	1. Infected fish scrapes itself against objects, presence of clamped fins, etc. 2. Parasites are visible (about 1/4 inches in diameter) on the body of the fish.	1. In the case of light infestations, the lice can be picked off directly using a forceps. 2. Bathing in 10 mg per litre of potassium permanganate solution for 10 to 30 minutes or treating the whole water tank with 2 mg per litre of potassium permanganate is also effective.
2	Black spot or lack Ick disease or diplopstomiasis	The disease is caused by flatworm larvae of the genus *Neascus* that	1. Infected fish scrapes itself against objects. 2. Development of small black	1. Treatment with uniodized salt (sodium chloride) at 1,000 to 2,000 ppm or $KMnO_4$ at

		burrows into the skin of a fish.	specks or smudges on the body and around the mouth.	approximately 2 ppm or at higher is very effective. In addition to this formalin or malachite green, or a combination of the two, copper, methylene blue and quinine hydrochloride also show good result. 2. The concentration of $CuSO_4$ treatments is calculated by dividing the culture water's total alkalinity by 100. For example, recommended $CuSO_4$ for culture water with a total alkalinity of 80 ppm is 0.8 ppm.
3	White spot disease or freshwater ich, or freshwater ick	This is caused by a ciliate called *Ichthyophthirius multifili*	1. Common behavioural changes are anorexia, rapid breathing, rubbing and scratching against objects, resting on the bottom, and upside-down swimming	1. Chemical treatments are similar to those employed for diplopstomiasis. 2. Heat treatment around 86^0F (30^0C) is also highly effective.

			near the surface, the eye becomes cloudy. 2. Development of several characteristic white spots on the body or fins of the fish.	
4	Costiasis	This is caused by a mastigophore called *Costia necatrix.*	1. Excess mucous production, respiratory distress, lethargy, clamped fins and resting on the bottom are the common signs. 2. Grey blue film appears on the skin, which often turns to red patches in later stages of infection.	Same as white spot disease
5	Whirling disease	Infection of myxosporean *Myxobolus cerebralis.*	1. Infected fish "whirl" forward in an awkward, corkscrew-like pattern instead of normal swimming. 2. Other symptoms include	In addition to those commonly employed in the treatment of white spot disease, some drugs, like furazolidone, furoxone, benomyl, fumagillin, proguanil and clamoxyquine, have been

			pancreatic necrosis, lesions and disintegration of the cartilaginous support of the organ of equilibrium.	shown to reduce the infection rates.
6	Red sore disease	It is caused by branched stalked ciliated protozoan *Heteropolaria colisarum*.	1. Common behavioural are anorexia, rapid breathing, rubbing and scratching against objects, resting on the bottom, upside down swimming near the surface, cloudy eye, etc. 2. Development of red coloured ulcers or cotton-like growth on the skin, scales and spine.	Same as white spot disease
7	Ergasilusiasis	Genus of copepod crustaceans *Ergasilus*, commonly called gill lice.	1. The fish scrapes itself against objects; whitish-green threads hang out of the fish's gills. 2. Red lesions often develop where the crustacean is	Same as white spot disease. But, the best treatment is bathing 10 to 30 minutes in 10 mg per litre of potassium permanganate solution.

			attached to the fish.	
8	Trichodiniasis	Disease is caused by the infection of *Trichodinia sp.*	1. Lethargy, scratching against any suitable object, breathing at the surface, or remain stationary at the surface are the most common behavioural symptoms. 2. Other common symptoms include presence of clamped or folded fins, darker body colours than normal, excess of mucus in gill and covering of skin by a pale bluish slime.	Same as white spot disease.
9	Anchor worms	Parasitic crustacean of the *Lernaea* species (copepod)	1. Fish infected with anchor worms have red and inflamed skin irritations. In some cases, the parasite's body is sticking out with whitish-green threads like	1. Salt and chemical treatments are similar to those employed for treating white spot disease. In addition, modern antiparasitics such as disco-worm, fluke tabs, and clout

			strikes. 2. Infected fish shows general lethargy, frequent rubbing or "flashing" against objects, inflammation on the body of the fish and breathing difficulties. Ulcers may also appear.	may help. 2. In the case of aquarium, the best treatment is to remove them by hand and quarantine the tank.
10	Tapeworms (Cestoda) infection	Common species are Asian tape worm *Bothriocephalus acheilognathii*, *Polyonchobothrium clarias*, *Ligula intestinalis* etc.		1. Chemotherapeutic agents like di-n-butyltin oxide, dibutyltin dilurate, yomesan when applied in food at a rate of one litre per 70kg dry weight effectively relieved fish from cestode infection. 2. Treatment of ponds with insecticides to eliminate copepods hosts of the parasites has also been recommended.

5.2 EPIZOOTIC ULCERATIVE SYNDROME

Epizootic ulcerative syndrome (EUS) is an epizootic condition of both wild and farmed freshwater and estuarine finfish. The infection is characterised by the presence of typical penetrating hyphae surrounded by granulomatous inflammation. It is also commonly known as mycotic granulomatosis (MG) or red spot disease (RSD) or ulcerative mycosis (UM). In 2005, scientists proposed that EUS should be named as epizootic granulomatous aphanomycosis (EGA).

Fish skeleton muscle is the main target organ, which exhibits the major EUS clinical signs with mycotic granulomas. It infects many freshwater and brackish water fish species in the Asia-Pacific region and Australia. The disease is most commonly seen in the tropical and sub-tropical waters with low temperature and heavy rainfall.

Etiological agent: The most probable causative agent of EUS is the oomycete water mould *Aphanomyces invadans* or *Aphanomyces piscicida.* EUS was first reported in Japan in 1971 in farmed ayu (*Plecoglossus altivelis*). Later in 1972, it was reported in estuarine fish, particularly in grey mullet in eastern Australia.

Susceptible stages: Juvenile and young adults are usually the most susceptible life stages of the fish. Though the Indian major carp, catla, rohu and mrigal are naturally susceptible species, an experimental injection of *A. invadans* into the yearling life stage of IMC revealed that they are resistant to EUS (Pradhan et al. 2009).

Mechanism of infection: The fungus has two typical zoospore forms- primary and secondary zoospore. The primary zoospore is released to the tip of the sporangium where it forms a spore cluster. It quickly transforms into the free-swimming infective stage called secondary zoospore. Infection occurs when motile secondary spores in the water or other carriers/ vectors arc attached to the skin of the fish. The spores gradually penetrate the skin, germinate and form fungal filaments or hyphae. The hyphae invade widely into the surrounding skin and deeply into the underlying muscles which result in extensive ulceration and destruction of tissues.

Signs and Symptoms: When EUS starts spreading into a fish pond, high morbidity (>50%) and high mortality (>50%) might be observed, particularly in a long cold season with water temperatures in between 18 to 22^0C.

EUS is clinically characterised by the occurrence of large haemorrhagic or necrotic ulcerative lesions on the bases of the fins and other parts of the body. Progressive diagnostic symptoms of EUS are:

A] Behavioural signs:

1) Reduced appetite, loss of balance, flashing.
2) Swim with head out of the water, fish swimming near the surface and sinking to the bottom.
3) Floating lethargically, cork-screwing or air gulping.

B] Clinical signs:

1) At the initial stage of infection, fish develop red spots on the skin.
2) Appearance of small amber-coloured, red or grey lesions.
3) Expansion of lesions into large ulcers, loss of scales and formation of oedema.
4) Extensive erosions filled with necrotic tissue and mycelium. This is followed by the development of granulomas on the internal organs and death.
5) Confirmation of EUS requires histological demonstration of the typical granulomatous inflammation around invasive hyphae or the isolation of *Aphanomyces invadans* from the underlying muscle.

Remedial measures/ Control measures/ Fish health management: The effective treatment and control of EUS are slowed down by several major problems, particularly the disease's widespread occurrences. Some of the common control and preventive measures for EUS are as follows-

1) **Vaccination:** There is no protective vaccine available till today. However, immunisation of snakehead fish with a crude extract of the *A. invadans* can elicit the humoral immune response as detected by SDS-PAGE (sodium dodecyl sulphate polyacrylamide gel electrophoresis) and Western blot analysis (Thompson et al. 1997).
2) **Removal of infected fish:** In the case of mild infection, the infected fish should be moved into the high quality water, where they may recover.
3) **Chemical treatment:** Application of agricultural lime ($CaCO_3$ or $CaMg(CO_3)_2$), burnt lime (quicklime- CaO or slaked lime– $Ca(OH)_2$), salt ($NaCl$), potassium permanganate ($KMnO_4$), bleaching powder, malachite green, iodine, formalin at recommended doses are often very effective in reducing mortality.

4) **Immunostimulation:** Intraperitoneal injection of the immunostimulant, salar-bec (300 g kg–1 vitamin C, 150 g kg–1 vitamin B and trace quantities of vitamins B1, B2, B6 and B12) into snakehead fish has been found to increase serum inhibition of both germination and growth of the zoospore in vitro (Miles et al. 2001).
5) **Antibiotics:** Antibiotics like tetracycline, erythromycin, nalidixic acid, oxytetracycline at the recommended dose for a recommended period also shows highly effective results. The study reported that bacteria isolated from EUS-affected fishes in India are resistant to gentamycin and amoxicillin, while sensitive to oxytetracycline, chloramphenicol and nalidixic acid (Saha and Pal 2002).
6) **Stocking with resistant species:** Culture species like nile tilapia, milkfish and Chinese carp are resistant to EUS and could be cultured in areas where EUS is endemic.
7) **Good husbandry practices:** EUS outbreaks are associated with heavy rainfall and flood events, drop in temperature, low alkalinity and salinity, acidified run-off water from acid sulphate soil areas. Thus, good husbandry, surveillance and biosecurity measures are necessary to prevent the occurrence or reoccurrence of EUS. Following husbandry practices must be employed for preventing EUS occurrence.
   a) Routine disinfection of fish eggs and larvae though there is no report of the presence of *A. invadans* in fish eggs or larvae.
   b) Increasing salinity in holding waters may also prevent outbreaks of EUS. Liming of water and improvement of water quality along with the removal of infected fish, are often effective in reducing mortality in case of an outbreak.
   c) Dead fish, contaminated fishing gears/net and fish transport containers should be prevented from getting into uncontaminated fish ponds.
   d) Avoiding farming susceptible fish species during EUS season, particularly in rainfall period and low-temperature season.
   e) Good water quality management, regular monitoring of fish health, good record keeping, and good farm hygiene are among the other common good husbandry practices.
   f) Avoid transportation of infected fish seed or fish to the uninfected area.

5.3 IMMUNITY TO DISEASE

5.3.1 Immunity

Immunity is the ability of the body to protect against all types of foreign bodies like bacteria, virus, toxic substances, etc. which enter the body. It also includes adequate tolerance capability of the body to avoid allergy and autoimmune diseases. Thus, immunity keeps animals from being affected by a disease.

The immunity provided by our immune system is made up of special cells (basophil, dendritic cells, macrophages, mast cell, etc.), proteins (interferon, cytokines, interleukin, etc.), tissues (lymphoid tissue, Peyer's patches), and organs (spleen, thymus etc.) that fight with the infection and kill pathogens when they invade the body. The immune system "scans" the body to identify any substance that it considers as foreign. Thus, it discriminates between "self" and "nonself". There are three categories of immune protection- it can be

a) Innate or acquired,

b) Active or passive, and

c) Natural or artificial (Fig. 5.2).

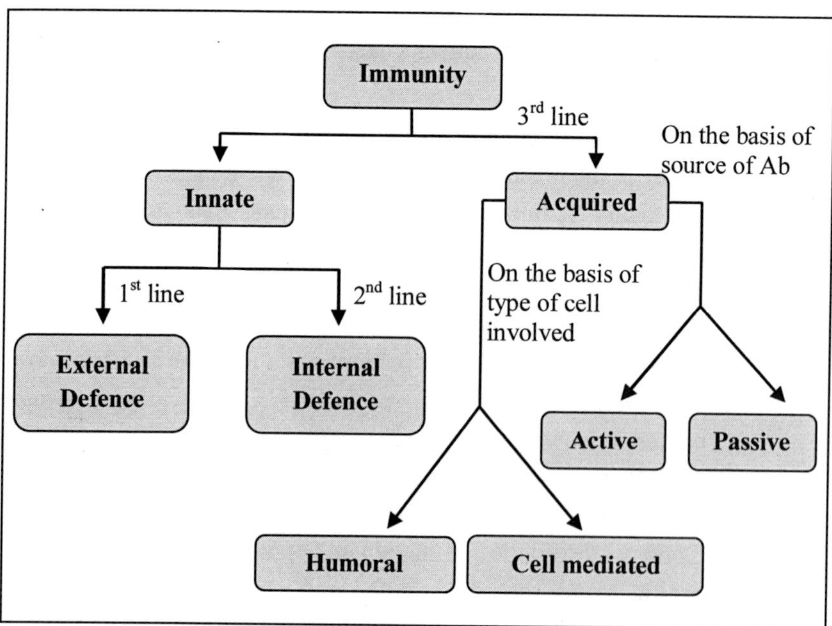

Fig. 5.2: Immune system of higher vertebrates involves both innate and adaptive immunity.

5.3.2 Immune response in fish

Fish are free-living aquatic organisms and are often prone to diseases, mainly due to a) increased stocking densities b) infected carriers c) infected facilities d) poor nutrition and e) substandard water quality. But all fish have mechanisms to protect themselves from a wide variety of infections or to fight diseases, although the immune system is not so highly advanced like the one that is found in mammals.

5.3.2.1 Cells and organs of immune response in fish

All fish possess immune cells like lymphocyte populations (similar to T cells and B cells) cytotoxic cells, macrophages and polymorphonuclear leukocytes. But, the immune organs vary by type of fish. In teleosts, the principal lymphoid organs are thymus, kidney (anterior and middle) and spleen. The anterior kidney (pronephros) in teleost fish is equivalent to the bone marrow in vertebrates. It is the largest site of haematopoiesis until adulthood, where erythrocytes, granulocytes, lymphocytes and macrophages develop. Scattered immune areas are also found in mucosal tissues in the skin, gills, gut and gonads.

The epigonal organs, the Leydig's organ found in the walls of the oesophagus, and a spiral valve found in the intestine are the three specialised lymphoid organs found in Chondrichthyes. In some fish (sturgeons, paddlefish, etc.) the mass that is associated with the meninges and heart also contains immune cells like lymphocytes, reticular cells and a small number of macrophages. Like bony fish, the chondrostean kidney is also an important haemopoietic organ.

5.3.2.2 Types of immune response

All metazoans possess an innate immune defence system, but the components of the adaptive immune system make their first appearance in the jawed vertebrates (gnathostomata). Thus, fish are the first animal phyla having the both innate and adaptive immune system (Magnadottir 2010).

5.3.2.2.1 Innate or nonspecific immune response

Innate immunity is defined as the natural immunity that is inherited by an organism from the parents and protects the organism from birth throughout life. It is also known as the non-specific immunity or inborn immunity. As its name nonspecific suggests it provides a non-specific defence against anything that is identified as foreign or non-self. It is the fundamental defence mechanism of fish. It is composed of the following physical barriers, immune cells and protein molecules.

1) **Physical barriers:** The first line of defence against infection is provided by scales, skin, skin mucus and gills. The mucus of fish contains many bactericides and fungicides like lectins, pentraxins, lysozymes, antibacterial peptides, complement proteins and immunoglobulin M (IgM), which kills or inhibits the entry of pathogens.
2) **pH of digestive tract:** Though there is a physical barrier, pathogens can still enter the fish's body, either through physical injury or the digestive tract. However, the digestive system contains antimicrobial active enzymes (lysozyme) and a very pathogen-unfriendly pH level.
3) **Antimicrobial peptides:** The low molecular weight antibacterial polypeptides found in the mucus, liver and gill tissue of teleost fish have the ability to kill bacteria by degrading their cell walls.
4) **Natural killer cells:** Nonspecific cytotoxic cells are widely found in lymphoid tissues of kidney and spleen. They are functionally similar to natural killer cells (NK cells) of mammals. The cytoplasm of these cells contains proteins such as perforin and proteases commonly known as granzymes. By perforin, NK cell creates pores in the plasma membrane of pathogen infected cell, allowing entry of granzymes and other molecules. This results in the killing of intracellular pathogens by osmotic cell lysis.
5) **Phagocytosis:** It is one of the prime innate defence system found in poikilothermic animals as it is least influenced by temperature. The main phagocytic cells found in fish are neutrophils and macrophages. These cells kill the engulfed pathogen either by lysosomal hydrolytic enzymes or by producing reactive oxygen species, myeloperoxidase, halide, hydrogen peroxide, nitric oxide and peroxynitrites, etc.
6) **Natural antibodies:** It has been found that in the absence of antigenic stimulation, many fish possess a high amount of natural antibodies like IgM in serum. They both provide immediate protection against bacterial and viral pathogens and stimulate the adaptive immunity.
7) **Other protein molecules:** In addition to the above nonspecific cellular and physical defence system, fish also possess many protein

molecules in their serum. They provide a second line of defence when the first line of defence fails to prevent access to pathogens in the body. For example,

a) Fish serum contains tumor necrosis factor (TNF) like TNF-α and -β which causes the activation of macrophages.
b) Interferons are glycoproteins released by virus-infected cells. Interferon makes nearby healthy cells resistant to viral attack by interfering with viral replication. IFNs also activate innate immune cells like NK cells and macrophages. The presence of interferons like INF1, INF2, INF3, INFα-1 and INFα-2 has been reported in rainbow trout and Atlantic salmon (Zou et al. 2007; Sun et al. 2007).
c) Interleukin, like IL-1β is found in many teleost fish species. It induces macrophage activation, generates inflammatory and chemotaxis.
d) Many fish possess protease inhibitors (e.g., α-2 macroglobulin) in the serum and other body fluids which degrade the proteolytic enzymes of pathogens.
e) C-reactive protein (CRP), serum amyloid protein (SAA) and transferrin have been reported in several teleost fish species, including rainbow trout, catfish, Atlantic salmon, common cod, halibut and dogfish. These proteins play an important role in activation of the classical complement pathway and the removal of apoptotic cells. Transferrin is an iron chelator. Thus, transferrin prevents infection as iron is essential for the establishment of infection by many pathogens.

It has been found that serum C-reactive protein (CRP) levels dramatically increases in rainbow trout when they are exposed to formalin, metriphonate which are used in aquaculture as anti-ectoparasitic chemicals. A significantly higher CRP level is found in fish reared at a water temperature of 16.5-19.5^{0}C compared to those reared at 13^{0}C (Kodama et al. 2004).

5.3.2.2.2 Acquired or specific immune response

Immunity that an organism developed against a specific pathogen or antigen when a person's immune system encounters foreign substances in the course of its lifespan is called acquired immunity. It is also called specific immunity as it is highly specific to a particular pathogen. Adaptive immunity is the 3rd line of defence developed by an animal in response to an invader.

Like mammals, the adaptive immune system in fish is composed of molecules corresponding to MHC antigens, T-cell receptors, B-cell receptors, antibodies, and a large number of accessory molecules. They respond to antigens in a highly specific manner.

1) **Humoral immunity:** Fish possess B-cells, which are structurally similar to mammalian B cells. Surface IgM of B-cells serves as a receptor for antigen recognition. Unlike other vertebrates, IgM in teleost fish is a tetramer. In some teleosts, it is a monomer. The second immunoglobulin isotype identified in fish, specifically in catfish is IgD. Recently, a unique isotype IgT has been found in rainbow trout (Hansen et al. 2005). In teleost, antibodies are predominantly found in the skin, intestine, gill mucus, bile and in the plasma.
2) **Cell mediated immunity:** The exact mechanism and the type of cells that are responsible for cell-mediated cytotoxicity in fish are poorly understood. However, several studies indicate that the mechanism of T cell mediated cell death also exists in fish. Antigen presentation by MHC is similar to that which occurs in higher vertebrates. The CD8+ T cell can recognise and bind to MHC class I-antigenic peptide complex. The activated cytotoxic T cell cells can induce apoptosis/ cell lysis and DNA fragmentation of the infected cells.

5.3.2.3 Factors influencing immune response of fish

The immune system and response of fish are greatly influenced by various internal and external factors. Some of them are briefly described below:

1) **Temperature and photoperiod:** Decreases in temperature, affect the rate of physiological functions in fish. Thus, the immune response is weaker at cold temperatures. For example, carps become immunologically suppressed at temperatures below 15°C, while coldwater species like salmon, shows suppression when the surrounding temperature below 6°C.
2) **Photoperiod:** An increase in daylight hours causes the reduction in the number of circulating leukocytes, while an increase in the

photoperiod causes an increase in the activity of lysozyme and circulating levels of IgM.

3) **Oxygen level:** Lower respiratory burst activity of macrophages and the levels of circulating antibodies are found to be associated with hypoxia or low dissolved oxygen in the water.
4) **Salinity:** Activity of different lytic enzymes, the respiratory burst activity of macrophages and levels of circulating IgM increases with an increase in the salinity.
5) **Stress:** Stress in fish due to high stocking density increases the circulating cortisol levels, while decrease the specific and non-specific immunity. Thus, stress makes the fish more prone to opportunistic infections.
6) **Nutrition:** Feed quality and quantity, nutrient availability, use of immunostimulants (e.g., β-glucans), probiotics (e.g., Lactobacillus, photosynthetic bacteria or yeasts,) in feeds improves the fish immune response against pathogens.
7) **Fish:** Immune response also depends on the age and species of fish. For example, rainbow trout are immune competent at an early age of their development.

5.4 FISH SPOILAGE

5.4.1 Spoilage

Spoilage is the undesirable process or post-harvest changes in food in which either the quality of edibility becomes reduced or it deteriorates to a point in which it is no more edible to humans. The spoilage of fish starts as soon as the fish dies. Characteristically a spoiled fish generally

1) Have a fishy off-flavour and ammonia-like strong and very unpleasant smell.
2) Appear to be dry or mushy in certain areas.
3) Have slimy gills.
4) Have flesh that is soft, or does not spring back when pressed upon and
5) Have a green or yellowish discoloration.

5.4.2 Causes of fish spoilage

Water, protein and fat are the main components of fish. Spoilage is brought about by the actions of enzymes, bacteria and chemical constituents present in the fish.

1) **Enzymatic action:** It is responsible for the early loss of quality. Some of the common enzymatic changes in fish muscles are:
   a) As soon as a fish dies, blood circulation stops, and thus the oxygen supply to the tissue are prevented. As a result post mortem glycolysis or anaerobic respiration occurs in muscle which converts glycogen into lactic acid. Lactic acid accumulation lowers the pH of the muscle. Lower muscle pH makes the fish muscle more susceptible to microbial attack (Fig. 5.3).

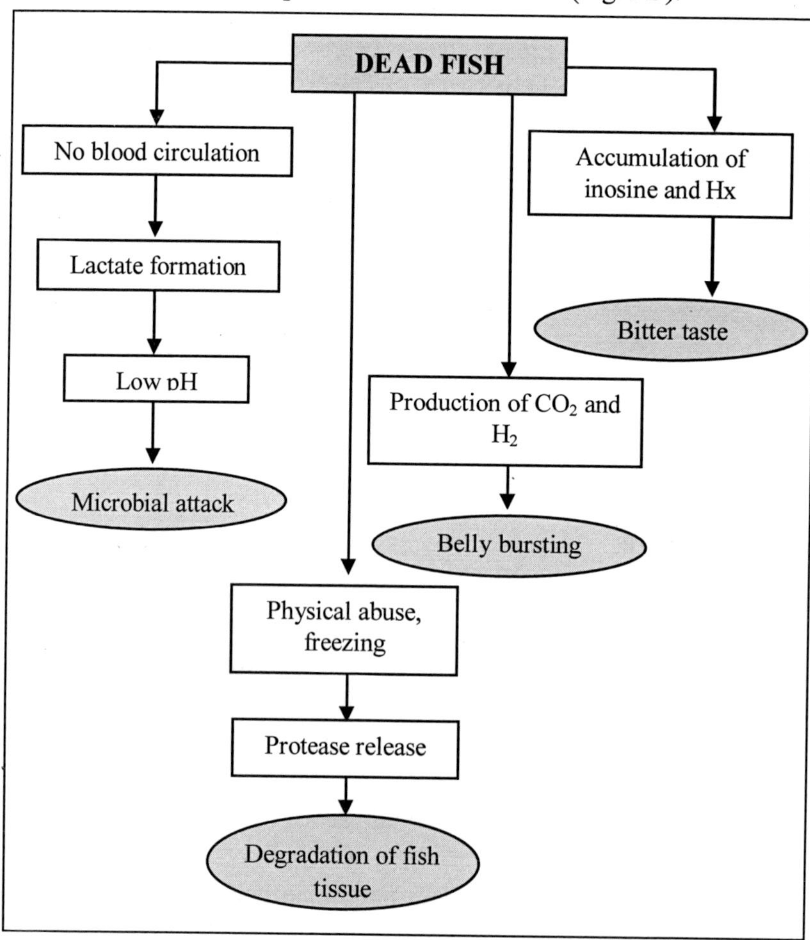

Fig. 5.3: Fish spoilage due to enzymatic degradation.

b) Due to enzymatic autolysis, the level of IMP (inosine monophosphate) decreases, while the level of neutral tasting inosine or bitter tasting hypoxanthine increases in tissue. This imparts a bitter taste in the fish muscle.

c) After death, digestive enzymes present in the gut dissolve gut components. Bacteria can proliferate in the dissolved gut contents very rapidly and produce gases such as CO_2, and H_2. This gas production often leads to belly bursting after a short storage period.

d) Proteolytic breakdown due to muscle proteases (e.g., carboxy-peptidases A and B, chymotrypsin, and trypsin) is the cause of extensive softening of the tissue. Acid proteases like cathepsins, calpains, collagenases are generally released into the cell juices due to physical abuse or freezing and thawing of postmortem muscle. These enzymes play a major role in the autolytic degradation of fish tissue (Table 5.2).

Table 5.2: Different autolytic changes in fish during spoilage (Huss 1995).

No	Substrate	Enzyme	Diagnostic changes
1	Glycogen	Glycolytic enzymes	Production of lactic acid, pH of tissue drops, loss of water holding capacity in muscle, high temperature rigour may result in gaping.
2	ATP, ADP AMP and IMP	The autolytic enzymes are also involved in nucleotide breakdown.	Loss of fresh fish flavour and gradual production of bitterness due to hypoxanthine formation at the later stage of spoilage.
3	Proteins and peptides	Cathepsins	Softening of tissue which makes processing difficult or impossible.
4	Proteins and peptides	Chymotrypsin, trypsin, carboxy-peptidases, etc.	Autolysis of the visceral cavity (belly- bursting).

5	Myofibrillar proteins	Calpain	Softening or molt-induced softening occurring mainly in crustaceans.
6	Connective tissue	Collagenases	Gaping of fillets and softening.
7	TMAO	trimethylamine-N-oxide (TMAO) demethylase	Formaldehyde-induced toughening of frozen gadoid fish.

2) **Microbial action:** After death, bacteria that are present in the skin mucous, gills and intestine of fish, invade the tissues and causes spoilage by the production of many undesirable compounds. Some of the common kinds of microbial fish spoilages are:
   a) The free amino acids in the fish muscle are delaminated by bacteria into toxic ammonia.
   b) Bacterial species of the genera *Shewanella*, *Pseudomona, Photobacterium* and *Moraxella* convert odourless compound called Trimethlyamine oxide (TMAO) into foul smelling trimethlyamine (TMA).
   c) Spoilage organisms like *Morganella morganii, Hafnia alvei*, etc. causes the formation of primary amines like histamine from histidine, arginine and glutamic acid.
   d) Decomposition of water-soluble nitrogen containing compounds like free amino acids and nucleotides into hydrogen sulphide (H_2S), dimethyl sulphide $(CH_3)_2S$ and methyl mercaptan (CH_3SH). These compounds are responsible for off-odors and off-flavors in spoiled fish (Table 5.3).

Table 5.3: Different compounds formed during bacterial spoilage in fish (Huss 1995).

No	Substrate	Compounds produced by decomposition
1	Inosine	Hypoxanthine
2	Carbohydrate and lactate	Acetic acid, CO_2 & H_20
3	Methionine & Cysteine	H_2S, CH_3SH and $(CH_3)_2S$

4	Tryptophan	Indole
5	Glycine, Leucine & Serine	Esters of acetic, Propionic, butyric and hexaenoic acids
6	Trimethylamine oxide	Trimethylamine
7	Urea	Ammonia
8	Lipids	Carbonyls
9	Proteins	Tyrosine, Indole, skatole, putrescine, cadaverine
10	Histidine	Histamine

3) **Chemical action:** Fish contains a high level of polyunsaturated fatty acids (PUFA). Lipid oxidation processes

- Auto-oxidation (action of O_2 on the unsaturated fatty acids) and
- Lipid hydrolysis (enzymatic hydrolysis of free fatty acids) produces aldehydes, ketones, alcohols, small carboxylic acids, fatty acid hydroperoxides and alkanes.

These are responsible for rancid flavour, odour and discoloration of spoiled fish. In addition, the fatty acids themselves may cause a "soapy" off-flavour in preserved fish.

5.4.3 Factors affecting fish spoilage

Spoilage is caused by the action of enzymes, bacteria and chemicals present in the fish. But, the rate of spoilage increases with high moisture content, high-fat content, high protein content, weak muscle tissue, ambient temperature and unhygienic handling. For example,

1) **Temperature:** In general, high temperatures increase the rate of fish spoilage and low temperatures slow it down.
2) **Careless handling:** Rough handling and bruising often speeding up the rate of spoilage as they cause contamination of fish flesh with bacteria and releases of autolytic enzymes. Additionally, careless handling can burst the guts and spread the contents into the fish flesh.
3) **Intrinsic factors:** As per the report of Food and Agriculture Organization (FAO) of the United Nations spoilage rate is higher in the round, small fish, fatty species and fish with thin skin compared to flat, large fish, lean species and fish with thick skin.

4) **Moisture content:** High moisture content is also responsible for fish spoilage. For example, seafood are less stable because of their high moisture content.

5.5 PROCESSING AND PRESERVATION OF HARVESTED FISH

5.5.1 Processing and preservation

Processing and preservation refers to all the techniques used to prevent fish spoilage due to microbial action or autolytic enzymatic damage and lengthen shelf life, while maintaining their flavour, taste, and nutritive value. The key principle behind the processing and preservation method is that it should be done in such a manner that the fishes remain fresh for a long time (long shelf life), with a minimum loss of flavour, taste, odour, nutritive value and the digestibility of their flesh.

5.5.2 Processing and preservation of harvested fish

Fish from the fresh and marine water bodies of the world have been a major source of perishable food, particularly protein for humankind since before recorded history. Thus, the retention of nutritional properties and product quality of fish is a very important part of the fishery industry. Rudimentary processing techniques like sun-drying, salting, and smoking were the common practices used by ancient Egyptians, Greeks, and other Mediterranean peoples to stabilise the fish supply. Processing and preservation is a very important part of commercial fisheries mainly because of the following two reasons:

1) Due to the distance between the fish farms or other fish capturing sites from the marketplace, there is always a chance of fish decomposition.
2) The uncertainties of sale in the market, particularly when the fishes are caught in large numbers which is greater than the amount of consumption. Thus, their preservation becomes a necessity for their future use.

Method of processing and preservation: The term fish processing refers to all the processes that are associated with fish from the time of it's harvesting, to the time of its delivery to the consumer. The method commonly used may be summarised as follows:

1) **Chilling or live fish trade:** Harvested fish must be immediately stored in a low-temperature environment in order to prevent the growth of microorganisms and inhibiting the enzyme activity. Refrigerated sea water cooling is advantageous over ice cooling because refrigerated seawater cooling causes less bruising and other structural damages to the fish carcasses compare to ice cooling. However, fish cooled in refrigerated seawater absorb salt from the water. Thus, care should be taken at the time of seawater cooling or it should be adjusted during the salt addition step of canning or smoking processes.

   On the other hand, in live food fish trade, fish are placed in a container with clean water with low temperature and optimum oxygen level and directly delivered to the market or consumers. Live fish can be transported either by simple traditional methods where the fish are placed in plastic bags with an oxygenated atmosphere or by sophisticated systems which use trucks that filter and recycle the water and always maintain optimum oxygen and temperature (Lima dos Santos et al. 1981).

2) **Preprocessing:** This step prepares the raw material (fish) for final processing. It is often performed on shipboard or in a shore-based plant. It includes different operations like inspection, washing, sorting, grading, and butchering of the harvested fish. In butchering after cleaning, the fishes are cut along their midventral side, and their visceral organs and other non-edible portions such as head, tail, and fins are removed.

3) **Final processing of fish (preservation):** Food preservation techniques include processes that control temperature, water activity, microbial loads, oxygen reduction potential, etc. Some of the traditional and modern or industrial techniques used for fish preservation are briefly discussed below:

   A] Control of temperature dependent spoilage: One of the main reasons behind the fish spoilage is metabolic activity in the fish occurring from microbial or autolytic processes. This can be reduced or stopped completely by decreasing the temperature. The three common methods used for controlling the temperature are:

   1) **Refrigeration:** It is simply cooling where the temperature is dropped to about 0°C and is generally used for short duration

storage. The idea behind refrigeration of the dead fish is to retard spoilage by slowing down the bacterial action. On fishing vessels, the fish can be refrigerated mechanically either by circulating cold air or by packing the fish in boxes with ice.

2) **Freezing:** The idea behind freezing is to completely stop the spoilage by stopping bacterial action and autolytic enzymatic damage. In freezing, the temperature is dropped below -18°C. In rapid freezing, fishes are quickly frozen from 0 to -18°C. Rapid freezing is advantageous over slow freezing as the former allows smaller sized ice crystals formation, takes shorter time for freezing and require less time for the diffusion of salts and evaporation of water and also prevents decomposition. Fish which are frozen properly at -20°C retains its physical properties and nutritive values for a year or more and is almost as good as fresh fish.

3) **Pumpable ice technology:** It produces cooling medium or slurry ice crystals (5 to 10,000 micrometres in size) called pumpable ice (PI) or coolant with a similar viscosity of water and the cooling capacity of ice. The advantages of PI over freshwater solid ice are- a) it is homogeneous and cooling rate is higher and b) it flows like water, and thus chances of freeze burns and physical damage to the cooled object is much lower.

B] Control of water dependent spoilage: Presence of water in fish tissue is necessary for the microbial and enzymatic reactions involved in spoilage. The common techniques that are generally used to remove the water from fish are:

1) **Smoking:** Originally smoking was done as a preservative, but now a day it is generally done for the unique taste and flavour developed in fish in the smoking process. Smoking is one of the oldest preservation methods which combine the effects of salting, drying, heating and smoking.

In this method, butchered fish is first brined in a 70 to 80 percent brine solution for a few hours to overnight. The fish are then partially dried on racks. It is then exposed to cold or hot smoke treatment. In cold smoking, first, a temperature of 38^0C is raised from a smokeless fire. After this heating step, cold smoke (< 28^0C) is allowed to circulate over the fish. In the case of hot smoking, first, a temperature of 130^0C is raised, followed by circulation of hot smoke (40^0C) over the fish.

2) **Drying:** Drying is one of the oldest techniques in which fish is dehydrated or desiccated. Thus, drying inhibits the growth of bacteria, yeasts, and mould and enzymatic autolysis. Drying can be done either by the traditional methods like air drying, sun drying, smoking or wind drying or by modern methods like electric food dehydrators, vacuum drying or freeze-drying.

3) **Salting:** Salting is the preservation technique in which water is removed from a substance through a process called osmosis. Salt (20%) draws water out (dehydrated) of microbial cells through osmosis, and thus kill them (plasmolysis). During this process, small fishes can be directly salted without being cleaned, while in the case of medium and large sized fish like sharks, the head and viscera are removed and cut into convenient sized pieces before salting. Salting can be of the following three types:

   a) **Dry salting:** In this process, the fish is first sprinkled or rubbed generously with salt and then packed in layers in the tubs and cemented tanks. After 10-24 hours, the fishes are removed from the tubes, washed in a salt solution and dried in the sun for 2 or 3 days. In hard-cure salting, an additional step is used in which warm air passes over the fish until the water content is reduced to around 20 percent and the salt content is increased to around 30 percent.

   b) **Wet salting/ Pickle-curing:** In this method, the cleaned fish are put in the previously prepared brine (salt solution, 270g of salt per litre of water for strong brine or 120g of salt per litre for weak brine). This is used for fatty fish such as herring. The fishes in cans are then marketed without drying.

   c) **Mixed salting:** In the mixed salting process, simultaneous use of salt and brine is followed.

4) **Freeze-drying:** In freeze drying, fish are dried in two stages, first by freezing the material (between −50°C and −80°C) below its triple point and then by reducing the surrounding pressure and raising the temperature. The latter allows sublimation of frozen water directly from the solid phase to the gas phase.

   A freeze drier generally consists of a vacuum chamber containing a) trays to hold fish during drying, b) source of heat to supply latent heat of sublimation, c) refrigeration coils to condense vapour and d) vacuum pump to remove non-expansible vapour.

5) **Water-binding humectants:** Humectants (a molecule with several hydrophilic groups) are additives that bind water and control water activity. Thus, humectants lower the rates of microbial degradation by lowering the water activities. Propylene glycol, sucrose, sorbitol, xylitol, maltitol and sodium chloride are the examples of humectants.

C] **Control of microbial spoilage:** Microbial growth and proliferation can be inhibited by the following preservation techniques:

1) **Canning:** It is a method of preservation in which fish spoilage is prevented by killing micro-organisms through heat. Canning provides a shelf life typically ranging from one to five years. It involves cooking the fish, sealing it in a sterile airtight cans or jars, and boiling the containers to kill or weaken any remaining bacteria. Bacteria like *Clostridium botulinum* needs a temperature of 120^0C for 4 minutes or at 115^0C for 10 minutes to kill them.

2) **Chemical control (preservatives):** Common chemical preservatives used in fish preservation include a) antimicrobial preservatives (prevent degradation of fish by bacteria) like nitrites, sulphites, sorbates, benzoates and essential oils, etc. and b) antioxidant preservatives (they prevent or inhibit the oxidation process) like butylated hydroxyanisole (BHA), butylated hydroxytoluene (BHT), ascorbic acid (vitamin C) and ascorbates, etc.

3) **Biopreservation:** Multiplication of bacteria is generally ceased when the pH drops below 4.5 (acidic). In biopreservation, natural microbiota or antimicrobial agents are used to lower the pH of fish muscle. The biopreservative agents must be harmless to humans and they must produce metabolites which is harmful to microbes. Different biopreservative agents are:

   a) **Lactic acid bacteria:** Metabolites of lactic acid bacteria (LAB) such as lactic and acetic acid, hydrogen peroxide, nisin, and peptide bacteriocins are active antimicrobials. Thus, during preservation, the addition of LAB, both lowers the pH, which unsuitable for microbial growth and produced antimicrobial agents which kill the microbes.

   b) **Yeast:** It has been reported that yeasts also have a biopreservation effect due to their competition for nutrients and capability for production and tolerance of high concentrations of ethanol. It also synthesises a large class of antimicrobial agents.

c) **Bacteriophages:** The bacteria eaters are viruses which infect bacteria. Phage preparations specific for the bacteria like *L. monocytogenes*, *E. coli* O157:H7 and *S. enteric* have been commercialised and approved for application in foods as biopreservative (Perez Pulido et al. 2015).

D] **Control of the oxygen reduction potential:** By reducing the oxygen content around fish its shelf life can be increased as both spoilage bacteria and lipid oxidation needs oxygen. It can be done by:

1) **Modifying environment:** In this method, surrounding oxygen (20.9% to 0%) is replaced by nitrogen (N_2), carbon dioxide (CO_2), carbon monoxide (CO). Low oxygen content, slows down the growth of aerobic organisms, while CO_2 lowers the pH, and thus inhibit the growth of bacteria.
2) **Vacuum packaging:** It involves complete removal of air from the package (plastic bags, canisters, bottles, etc.) prior to sealing using the vacuum chamber machine. It reduces atmospheric oxygen, limiting the growth of aerobic bacteria or fungi.

5.5.3 Advantages and disadvantages of fish preservation:

The preservation and processing is a very important aspect of the fish industry as it lengthens shelf life from days to a year or longer of harvested fish, while in general maintains their flavour, taste, and nutritive value. However, it has certain drawbacks such as:

1) The growth of botulinum bacteria in fermented fish kept in anaerobic conditions in the airtight container.
2) A slow freezing process often causes denaturation of flesh due to the formation of ice crystals in the muscles. As a result, the preserved fish becomes dehydrated, loses texture and often lose flavour and taste.
3) Incomplete or poor preservation often leads to decarboxylation of histidine of fish flesh into histamine. Histamine is one of the common causes of food poisoning.

4) Weight, nutritive value and the digestibility of the flesh is reduced in the drying process. Sometimes salt tolerant bacteria grow on highly salted fish which causes **pink eye spoilage** of fish flesh. It has been reported that salting combined with smoking results in loss of protein, about 1 to 5 % due to salting and 8 to 30 % due to smoking.
5) Smoking also accelerates rancidity of fat and so reduces digestibility of fat products. Canning leads to much loss of vitamin B1, pantothenic acid, vitamin-C and pteroylglutamic acid.
6) Preservatives like sodium nitrate increase the likelihood of occurrence of pancreatic and lung cancer, propyl gallate can cause prostate inflammation and tumours in the brain, pancreas and thyroid. Antioxidant preservatives like butylated hydroxyanisole (BHA) and butylated hydroxytoluene (BHT) are used to prevent oxygenation. This keeps fats from turning rancid, and thus extends the shelf life of any food containing fat. Though the harmful effects of BHT are still unproven, but BHA has been found to cause cancer in animals (Branen 1975).

5.6 FISHERY BY-PRODUCTS:

The fish represents a valuable source of proteins and nutrients, and thus the most common use of fisheries resources is food. In addition to food, the fish industry also yields a number of byproducts like body oil, liver oil, fish emulsion, fish glue, isinglass, fancy articles and many more for commercial purposes. Some of the by-products of the fishing industry are as follows:

1) **Surimi:** It is a paste made from fish. According to the United States department of agriculture national nutrient database (USDA National Nutrient Database) fish surimi contains about 76% water, 15% protein, 6.85% carbohydrate, and 0.9% fat. Surimi curing is prepared by heating during which polymerization of myosin occurs. Fish with higher fat contents (lack heat-curing myosin) are not used for surimi preparation. In some cases, the addition of egg white or potato starch is required to form firm surimi. Surimi is either eaten directly as a food or it is used to imitate other processed foods.
2) **Fish glue:** It is adhesive that is prepared by prolonged boiling of skin or bones of fish like cod, haddock, pollak and hake. Adhesives

made from fish were sometimes used in ancient Egypt for wood furnishings and mural paintings. They are still used in art, for shoe and furniture repair, and to preserve old manuscripts.

3) **Isinglass:** It is a pure, transparent or translucent form of collagen-rich gelatin, prepared from the air bladders of certain fish, particularly the sturgeon. It is widely used in glue and jellies and as a clarifying agent during clarification or fining of some beer and wine. During the brewing process, any particles and yeast cells floating in a beer stick to isinglass, congeal into a jellylike mass and sinks to the bottom of the cask, so that beer makers can easily remove it.
4) **Fish oil:** It is prepared from the tissues of oily fish like sardines, herring, salmon, trout, tuna and mackerel. Fish oils contain the omega-3 fatty acids eicosapentaenoic acid (EPA) and docosahexaenoic acid (DHA), which are precursors of certain eicosanoids. The eicosanoids are known to reduce inflammation in the body and have other health benefits like lower the risk of heart attacks or strokes and lower blood pressure.

   Some studies found that better psychomotor development occurs in infants whose mothers received fish oil for the first four months of lactation (Jensen et al. 2010). Patients taking omega-3 supplements show less depressive symptoms in depressive patients.
5) **Liver oil:** It is extracted from the liver of fish like Atlantic cod (*Gadus morhua*). It is one of the best sources of omega 3 fatty acids (EPA and DHA), vitamin D and E. The medical use of cod liver goes back to centuries. It contributes to the maintenance of muscle function, supports the maintenance of normal bones, contributes to the normal function of the immune system, support normal heart function, and supports the maintenance of normal vision and brain function (high cognitive performance). Thus, it prevents the risk of coronary atherosclerosis, age-related macular degeneration, pain relief in rheumatoid arthritis patients and helps in repairing wounds.
6) **Fish fertiliser:** It is widely used for promoting plant growth. It contains high contents of N, P and Ca. Fish fertiliser is made either from whole fish or carcass products, including bones, scales and skin. Fish fertiliser can be prepared directly by drying the fish in the sunlight. It can also prepared by burning the fish waste, followed by fermentation of the ash in a soil pit.

In most advanced method, it is produced from the remaining of fish meal. In this method, fish, such as menhaden and anchovies, are ground into slurry. This product is then processed to remove oils and fish meal, which are used in other industries. The liquid that remains after processing is the fish emulsion. After straining out solids, sulfuric acid is added to lower the pH. The fish fertiliser thus formed is stable and can be directly used in the garden.

7) **Fish meal:** It is mostly prepared from fish (e.g., anchovy, horse mackerel, pout, sandeel, sprat etc.) that are not generally used for human consumption. It is made by cooking, pressing, drying, and grinding of fish or fish waste. High-quality fishmeal normally contains 60-72% crude protein by weight. It also contains calcium, phospholipids, iodine and minerals. Fish meal is used as a protein supplement in the compound feed of fish, pig, poultry, cattle, etc.
8) **Fish sauce:** It is an amber-colored liquid. It is extracted from the fermentation of fish with sea salt. It is used in Southeast Asian cuisine much like salt or soy sauce as a flavour enhancer and commonly eaten with fish, shrimp, pork, and chicken.
9) **Fish hydrolysate:** It is nitrogen rich byproduct of fishery commonly used for promoting plant growth. The difference between fish emulsion & fish hydrolysate is the fish oil. Fish Hydrolysate still contains the oil and is undiluted and there are many uses of fish hydrolysate like-
   - It is a richer food source for beneficial microbes and especially beneficial fungi in the soil. Thus, it can be used as fish-based fertiliser.
   - Fish protein hydrolysate prepared from salmon contains significant cancer growth inhibitors.
   - Fish hydrolysates prepared from sardines (family-Clupeidae) have gut health and anti-hypertension benefits.
10) **Others:** Squalene (30-carbon organic compound) found in the liver oil of shark is used as a mordant in the dyeing of synthetic fibres. Sharks reduce their body density by storing fats and oils in their liver as they lack the hydrostatic organ swim bladder. Squalene is lighter than water with a specific gravity of 0.855. Similarly, lecithin found in the liver oil of dogfish and shark is used as a wetting in the chocolate industry. In addition to this, fish byproducts are traditionally used in the Ayurvedic and Unani system of medicine

for treating ulcers, skin diseases, cough, colds, bronchitis, asthma etc. (Table 5.4).

Table 5.4: Medicinal uses of animals and animal parts in traditional therapy (Vats and Thomas 2015; Teronpi et al. 2012).

Sr. no.	Name of fish	Parts used	Disease treated	Preparation	Dosage
1	Channa punctatus	Eyes	Corn or clavus disease	Eyes mixed with common salt	Applied to the infected part externally
2	Elephant snout fish (*M. kannume*)	Whole fish	Hookworm infection	Burn, grind, mix with hot water	1 cup/day for 3 days
3	Nile Perch (*Lates niloticus*)	Gills	Abdominal cramp	Pound and mix with water	1 cup/day for 7 days
4	Victoria tilapia (*O. variabilis*)	Scales	Cough	Burn and swallow the ashes	Regularly
5	Hammer head shark (*Zygaena blochii*)	Fat	Joint pain	Extracted from liver or tissue	Fat is applied externally
6	*W. attu*, *L. pangusia*, *Puntius sp.*, etc.	Whole fish	Weakness after delivery	Boiled fish	Regularly
7	*D. aequipinnatus*	Whole fish	Constant spitting	Boiled fish	Regularly
8	*Monopterus cuchia*	Whole fish, raw blood	Anaemia	Raw blood	Raw blood taken orally
9	Any hill stream fish	Belly	Sore festering wound	Dissect out belly	Bandage with fish belly for 2-3 days

Besides this, several compounds extracted from fish are also employed as component in many official medicines (Costa-Neto 2005). For example,

- Oily fish, like cod, herring, salmon, and turbot are well known sources of PUFA known as OMEGA-3 which helps the prevention of arthritis.

- Atlantic salmon and rainbow trout are important sources of an anticoagulant, similar to protein C anticoagulant found in mammals.
- Tetrodotoxin (TTX) isolated from puffer fish functionally resembles procaine and it acts as an extraordinary narcotic and analgesic.
- Fish like *Eptatretus stoutii*, *Dasyatis sabina*, and *Taricha sp*. are well known sources of cardiac stimulants, antitumors, and analgesic, respectively.

5.7 REFERENCES

Branen AL (1975) Toxicology and biochemistry of butylated hydroxyanisole and butylated hydroxytoluene. Journal of the American Oil Chemists' Society. 52: 59.

Costa-Neto EM (2005) Animal-based medicines: biological prospection and the sustainable use of zootherapeutic resources. Anais da Academia Brasileira de Ciencias 77 (1): 33-43.

Gopalakrishnan V (1963) Controlling pests and diseases of cultured fishes. Indian livestock. 1(1): 51-54.

Hansen JD, Landis ED, Phillips RB (2005) Discovery of a unique Ig heavy-chain isotype (IgT) in rainbow trout: Implications for a distinctive B cell developmental pathway in teleost fish Proceedings of the National Academy of Sciences of USA. 102(19): 6919–6924.

Hora SL and Pillay TVR (1962) Handbook on fish culture in the Indo-Pacific region. FAO Fish Biology Technical Paper No. 14, FAO, Rome, Italy, 204p.

Huss HH (1995) Quality and quality changes in fresh fish. FAO Fisheries Technical Paper No. 348, FAO, Rome, Italy, 195p.

Jensen CL, Maude M, Anderson RE, Heird WC (2000) Effect of docosahexaenoic acid supplementation of lactating women on the fatty acid composition of breast milk lipids and maternal and infant plasma phospholipids. American Journal of Clinical Nutrition. 71(1 Suppl): 292S-9S.

Jensen MH (1963) Preparation of fish tissue cultures for virus research. Bulletin de l'Office International des Epizooties. 59: 131–134.

Kodama H, Matsuoka Y, Tanaka Y, Liu Y, Iwasaki T, Watarai S (2004) Changes of C-reactive protein levels in rainbow trout (*Oncorhynchus mykiss*) sera after exposure to anti-ectoparasitic chemicals used in aquaculture. Fish and Shellfish Immunology. 16(5):589-597.

Lima dos Santos CAM, James D, Teutscher F (1981) Guidelines for chilled fish storage experiments. FAO Fisheries Technical paper No. 210, FAO, Rome, Italy, 22p.

Magnadottir B (2010) Immunological control of fish diseases. Journal of Marine Biotechnology 12(4): 361–379

Miles DJC, Polchana J, Lilley JH, Kanchanakhan S, Thomp-son KD, Adams A (2001) Immunostimulation of striped snakehead *Channa striata* against epizootic ulcerative syndrome. Aquaculture 195(1-2): 1–19.

Noga EJ (2010) Fish disease: diagnosis and treatment, 2nd edn. Wiley-Blackwell, Hoboken, New Jersey, pp.1-519.

Perez Pulido R, Grande Burgos MJ, Galvez A, Lucas Lopez R (2016) Application of bacteriophages in post-harvest control of human pathogenic and food spoiling bacteria. Critical Reviews in Biotechnology. 36(5): 851-861.

Pradhan PK, Mohan CV, Shankar KM, Mohana BK (2008) Infection experiments with *Aphanomyces invadans* in advanced fingerlings of four different carp species. In: Bondad-Reantaso MG, Mohan CV, Crumlish M, Subasinghe RP eds. Diseases in Asian Aquaculture VI. Fish Health Section, Asian Fisheries Society, Manila, Philippines, pp. 105-114.

Saha D, Pal J (2002) In vitro antibiotic susceptibility of bacteria isolated from EUS-affected fishes in India. Letters in Applied Microbiology. 34(5):311-316.

Sun B, Skjaeveland I, Svingerud T, Zou J, Jorgensen J, Robertsen B (2011) Antiviral activity of salmonid gamma interferon against infectious pancreatic necrosis virus and salmonid alphavirus and its dependency on type I interferon. Journal of Virology. 85(17): 9188-9198.

Teronpi V, Singh HT, Tamuli AK, Teron R (2012) Ethnozoology of the Karbis of Assam, India: Use of ichthyofauna in traditional health-care practices. Ancient Science of Life. 32(2): 99-103.

Thompson KD, Lilley JH, Chinabut S, Adams A, (1997) The antibody response of snakehead, *Channa striata* Bloch, to *Aphanomyces invaderis*. Fish and Shellfish Immunology. 7(5):349-353.

Vats R and Thomas S (2015) A study on use of animals as traditional medicine by Sukuma Tribe of Busega District in North-western Tanzania. Journal of Ethnobiology and Ethnomedicine. 11: 38.

Zhou Z, Hamming OJ, Ank N, Paludan SR, Nielsen AL, Hartmann R (2007) Type III interferon (IFN) induces a type I IFN-like response in a restricted subset of cells through signaling pathways involving both the Jak-STAT pathway and the mitogen-activated protein kinases. Journal of Virology. 81(14): 7749-7758.

*** ***** ***

Chapter 6

AQUACULTURE BIOTECHNOLOGY

Joyobrato Nath

Department of Zoology, Cachar College, Silchar, Assam, India

Chapter highlights:

In the last 3 to 4 decades, spectacular growth took place in the aquaculture industry. The vibrancy of the fishery industry in India can be visualised by the 11–fold increase in fish production in just six decades, i.e. from 0.75 million tonnes in 1950-51 to 9.6 million tonnes during 2012–13. Biotechnology provides powerful tools not only to meet high consumer demand, but also to improve aquaculture sustainably. Stock improvement through androgenesis, gynogenesis, transgenesis, hybridization, polyploidization help in the generation of positive traits like better somatic growth rate, hybrid vigour, reduction of unwanted reproduction through the production of sterile fish, improvement of disease resistance and cold resistance in fish, better fecundity, better food conversion capability, etc. In addition, induced breeding opens up a new era for breeding IMC in captivity. Long-term storage technique (cryopreservation) for preserving germ cells of threatened and endangered species for at least several thousands of years opens a new frontier for aquaculture research and conservation. Additionally, DNA barcoding using mitochondrial cytochrome c oxidase subunit 1 (CO1) offers the opportunity to identify phenotypically alike species, for identification and traceability of seafood, monitoring the trafficking of threatened species by fish poachers and many more.

6.1 SEX DETERMINATION IN FISH

Sex determination refers to the mechanism of establishment of either male or female sexual characteristics in the embryo. In general, the sex of the developing offspring depends either on the genotype of sperm (where the male is heterogametic sex) or ovum (where the female is heterogametic sex). In many species, sex determination is genetic, but in some animals, sex determination is under the control environmental variables such as temperature. Sexual development in fish is controlled by genetics as well as environmental factors.

1) **Environmental sex determination:** *Menidia menidia*, the first fish species where temperature-dependent sex determination (TSD) was found. High temperature usually stimulates more male production, while low temperature has either no effects or stimulates more female production. Till today, temperature-dependent sex determination has been reported in 59 species of fishes.
2) **Parthenogenesis:** It has been found that Amazon molly (*Poecilia formosa*), a freshwater fish reproduces through gynogenesis. The unfertilized egg develops into female parthenogenetically. As a result, Amazon molly becomes an all-female species animal.
3) **Chromosomal basis of sex determination:** In fish species, all kinds of sex-determining systems observed in other vertebrate classes have been observed, including male heterogametic (different types of gametes), female heterogametic.
   a) **Male heterogametic (XX-XY):** Several fishes like salmons, medaka (Japanese rice fish) etc. have heterogametic males and homogametic females, similar to the mammalian XX-XY system.

      In this determination system, males (AA+XY) are heterogametic sex and they produce two types of gametes (A+X) and (A+Y), while females (AA+XX) are homogametic sex and they produce only one type of gamete (A+X). Fertilisation of egg (A+X) with a sperm carrying the Y chromosome (A+Y), the zygote will develop into a male fish, while fertilisation of eggs with the sperm carrying the X chromosome (A+X), the zygote will develop into a female fish (Fig. 6.1).

      Thus, it is the genetic makeup of male gamete (sperm) which determines the sex of the offspring and there is always 50% probability (chance) of having either male or female fish in a particular mating.

b) **Female heterogametic (ZZ-ZW):** Fishes like Poecilia have homogametic males and heterogametic females (ZZ/ZW).

In this sex determination system, males (AA+ZZ) are homogametic and they produce only one type of gamete (A+Z), while females (AA+ZW) are heterogametic and produce two types of gametes (A+Z) and (A+W). Thus, it is the genetic makeup of female gamete which determines the sex of the offspring.

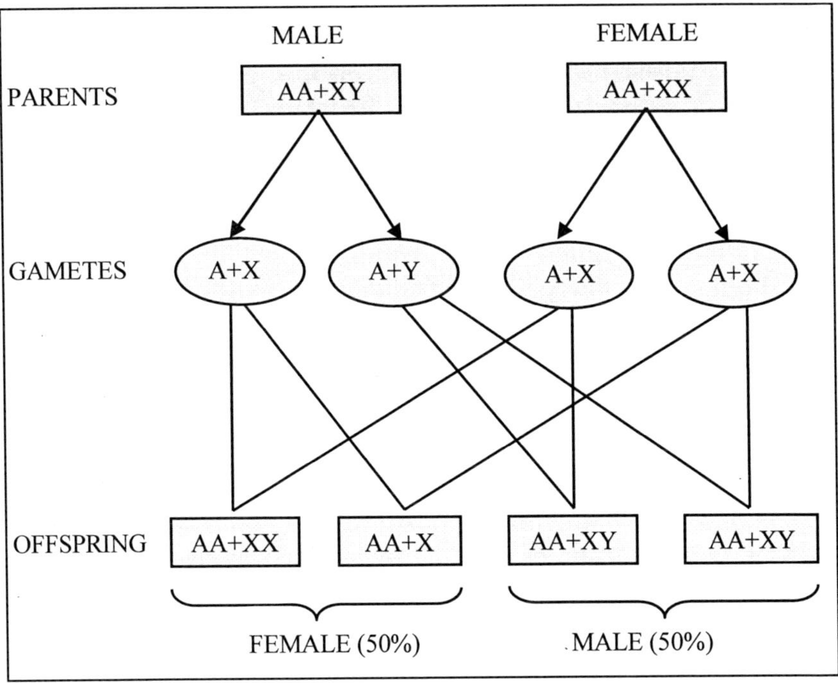

Fig. 6.1: XX-XY type of sex determination in fish.

c) **Three chromosome system (XX, XW, YW- XY and YY):** Some species of the platyfish *Xiphophorus*, utilise a system with three sex chromosomes. In the platyfish, three different sex chromosomes W, X, and Y are found. In these fish species, there are 3 types of females (XX, XW and YW) and 2 types of males (XY and YY). The W chromosome carries a stronger female determining potency than the X; because the WY constitution gives rise to females, while the XY constitution to give rise to males.

d) **Sex reversal in fish:** Though the exact mechanism is not known, but sex reversal (male to female or female to male) is found in many fish species in natural condition. For example, goby fish population generally contains a number of females and a single dominant male. With the loss of the male of the group, one of the adult female fish undergoes sex reversal to become the male of the group. Similarly, water temperature also influences sex reversal in adult zebrafish.

e) **Social variables based sex determination:** In some fishes sex determination and consequent differentiation is influenced by the social interaction. Social interaction, such as the disappearance of a male or female from a group causes sex inversion in these species either a) from male to female (anemonefish, *Amphiprion clarkii*) or b) from female to male (Red sea fish, *Anthias squamipinnis*) or c) either way or multiple times (serial sex change, goby fish, *Trima okinawae*) (Baroiller and D'Cotta 2001).

6.2 BREEDING OF FISH

6.2.1 Fish breeding

Fish breeding is the process by which offsprings are produced by sexually matured parent brood fish. Thus, breeding is necessary for species continuity and survival. In fishes, breeding is controlled by internal (endocrine system) and external factors (temperature, salinity, good water quality, adequate food and photoperiod).

6.2.2 Types of breeding

Breeding can be classified in different ways. On the basis of genetic similarity among breeds involved in breeding it can be classified into inbreeding and outbreeding, while on the basis stimuli involved, it can be classified into natural and induced or artificial breeding.

A) **Inbreeding:** Breeding of fish that are closely related genetically within the population is called inbreeding. Inbreeding produces offspring with lower levels of genetic variability or heterozygosity, and thus shows reduced biological fitness in a population called inbreeding depression.

B) **Line breeding:** Mating of individuals with its descendants is called line breeding. It increases an individual's contribution to the population and to succeeding generations. As the genetic contribution of an individual in a population is halved in each generation, after seven generations, a fish contribute less than 1% of its descendants' genes (Tave 1999). In this case, line breeding is can be used to reverse this trend.

C) **Outbreeding:** Breeding of fish from a different population or genetically unrelated individuals is called outbreeding. These genetic variants may have reduced biological fitness than their parents or progeny from crosses between individuals that are more closely related and is called outbreeding depression.

D) **Natural breeding:** Indistinct breeding of fishes in natural lotic or lentic habitats is called natural breeding. The river, swamp, bill, flooded land, outfall and coastal area are the major places of natural breeding of fish. In general, carp fish like rui, catla, mrigal etc. prefer to breed in small or big sized rivers, while taki, shol, koi, moda, common carp, etc. prefer to breed in the stagnant stream. Under favourable environmental condition, gonadotropin released from the pituitary gland incites the fish for breeding.

Monsoon season (April to July), cloudy sky, mild wind, a temperature in between 28°-31°C and a day with cold weather is very suitable for natural fish breeding. However, Tarpon breeds during the dry season, while tilapias breed throughout the year (Akankali et al. 2011). In general, during breeding season adult female fish swim towards or against the tide and the male fish follow it. In a little tidal area, the female fish lays eggs in the water and male fish release sperm over the eggs and fertilise the eggs. Naturally, the minnow comes out of the eggs within a few hours.

E) **Artificial breeding:** In the artificial method, breeding is carried out either by controlling the environment (bundh) or artificially inciting male and female fish (hypophysation). It is necessary for fish culture because of the following reasons:

1) Many cultural farm fishes like IMC do not breed in captivity. The reason may be environmental and hormonal. This problem can be solved by induced breeding
2) It assures timely availability of pure seed, whereas in nature the availability of seed is quite uncertain.
3) It can fulfil any quantity of demand at any time.
4) The cost of expenditure is lower than the natural collections of spawns.

6.2.3 Types of induced breeding

Fishes can be induced to breed either through a bundh called bundh breeding or hypophysation called hypophysation breeding.

a] Induced breeding through bundh: In bundh breeding, the IMCs have been induced to breed in special types of perennial and seasonal ponds locally called `Bundhs' in some parts of Bengal, Bihar and MP. The bundhs may be either perennial called wet bundhs or seasonal called dry bundhs. The large, shallow marginal areas of the bundhs serve as spawning grounds for the fish.

1) **Wet Bundhs:** A typical wet bundh is a perennial pond situated on the slope of a vast catchment area with an inlet towards the high land and an outlet at the opposite side towards the lower end to regulate the inflow and outflow of water respectively during heavy showers.

   It has an extensive shallow area which dries up during summer, but it also contains a deeper area which retains water throughout the year and where adequate stocks of brood fishes are maintained. After a heavy monsoon shower, fresh rainwater with washings from the upland area rushes into the bundh through the inflowing streamlets known as the "**dhals**".

   During heavy rains, a major portion of the bundh is submerged and the excess water is drained through the outlet called "**bulan**" which is guarded by bamboo fencing locally called "**chhera**". The flow of water through the outlet could be controlled by plugging the "**chhera**" with straw and mud. The shallow areas of the bundh, called "**moan**", are the main spawning ground of the carps (Kumar 1992).

   Breeding operation in wet bundh: With the onset of monsoon the fresh rainwater from the catchment area enters into the bundh. The excess water flows out from the bundh creating water current. The breeders present in the deeper area of the bundh migrate to shallow areas where they start breeding.

2) **Dry Bundhs:** Dry Bundh is a seasonal pond and is shallower than the wet bundh. With the first shower, when water accumulates, selected mature carps from the nearby ponds are released. Usually, with the next heavy shower, when the dry bundh is flooded, fishes get stimulated and breed.

   Thus, dry bundhs remain more or less dry during the greater part of the year. In a modified bundh, adjacent ponds are constructed along the gradient of the catchment area. The upper one where the pre-monsoon rainwater is collected from upland catchment area serves as a “reservoir” and the lower one is used for breeding purposes. The reservoir and breeding bundhs are arranged in a sequence along the gradient so as to facilitate the flow of water. The water flow is controlled by gates.

Breeding operation in dry bundh: Rainwater which accumulates in the catchment area during pre-monsoon showers flows in to fill up the pond seasonally. Thereafter, the brood fishes from a perennial pond are introduced into the seasonal ponds to breed, preferably on cold rainy days. Spawning usually commences during and after heavy showers when the bundh, as well as the catchment area, is flooded with fresh rainwater.

Factors essential for bundh breeding: The factors that are conducive to spawning in bundhs are: 1) water temperature of 27-29^{0}C at the time of spawning, 2) high turbidity, 3) low total alkalinity 4) little high conductivity of water and 5) abrupt rise of water level.

b] Induced breeding through hypophysation: Induced breeding is a technique in which the economically important fishes which generally do not breed in captive condition are breed through artificial stimulation. In induced breeding ripe fish breeders are stimulated by injecting the pituitary hormone called hypophysation technique.

Hypophysation breeding technique: The basic steps used in induced breeding through hypophysation are as follows-

1) **Choice of fish for the collection of the pituitary gland:** The gland should preferably be collected from fully ripe fishes, as the gland is most potent at the time of breeding or just before spawning. Most suitable time in India for the collection of the pituitary glands of major carps is during May to July months, as the majority of carps attain advanced stages of their maturity during this period.
2) **Collection of the pituitary gland:** The pituitary gland is situated on the ventral side of the brain just below the hypothalamus. The dorsal side of the head is first chopped with a knife. The dura mater covering the gland is then cautiously removed using a fine needle and forceps. The exposed gland is then picked up intact from the bony cavity (sella

turcica), without causing any damage to it. The gland can also be collected by dissecting through the foramen magnum.

3) **Preservation of gland:** Glands can be preserved in 100% ethyl alcohol. Acetone can also be used for preservation in temperate countries. The pituitary gland secretes the somatotropic hormone, adrenocarticotropic hormone (ACTH), follicle-stimulating hormone (FSH) and luteinizing hormone (LH).
4) **Preparation of gland Extract:** A known amount of the gland is taken by estimating the total quantity of fish to be bred. The gland is then taken in tissue homogenizer with a little amount of distilled water. The dilution rate is 0.2 ml/kg of body weight of the fish. The pituitary extract is then centrifuged and only the supernatant solution is used for injection. The extract can be preserved in pure glycerin for injection. The pituitary extract can also be stored in propane or 1.5% TCA (trichloroacetic acid) in a refrigerator for later use.
5) **Brooders selection:** The brooders must be healthy enough and ripe. 2 – 4 years of age with 1–5 kg body weight is preferable. The common selection criteria are-
    - The brood fish must be healthy enough and ripe.
    - It must be 2–4 years of age.
    - Generally, medium size with 1–5 kg body weight is preferable.
    - It must be free of disease and parasite.
    - They must be fed with artificial feed consisting of rice bran, oil cake, and powdered cotton seed daily for 15-30 days before the injection.
6) **Injection to the brooders:** The pituitary extract is administered into the body of brooder fish by means of hypodermic syringe either intramuscular or intra peritoneal in caudal peduncle or the shoulder regions near the base of the dorsal fin. While injecting the needle is first inserted under a scale parallel to the body of the fish and then it is injected quickly at an angle of 45^0. On the other hand, intraperitoneal injection is given at the base of pelvic or pectoral fin (Fig. 6.2).
7) **Dosage of the extract:** An initial dose at the rate of 2-3 mg/ kg body weight of fish is administered to the female brooder only. After an interval of time about 6hrs a second dose of 5-8mg/kg of body weight of the female. At the same time, a dose of 2-3 mg/kg body weight is given to the male.

8) **Spawning:** After injection to the brooder fish, a set of brooders (1 female + 2 male fish) is released into breeding hapa. The spawning takes place within 3-6hrs following the second dose. The fertilised eggs are transparent, pearl-like, whereas the unfertilised eggs are opaque or whitish.

Chinese carps however, do not spawn naturally and when they spawn, the percentage of fertilisation is generally very low. Stripping or artificial insemination is therefore followed. The eggs are collected in a plastic trough by pressing the belly portion of the female. Similarly, milt from the male is squeezed out into the same trough. The gametes are then mixed as soon as possible to allow fertilisation.

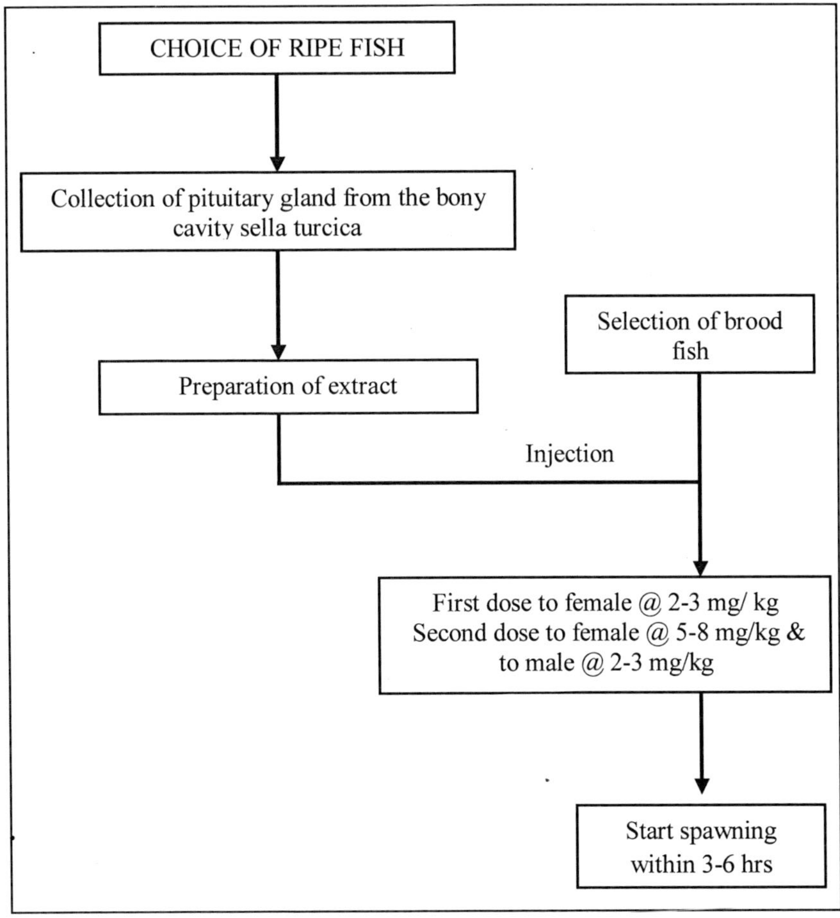

Fig. 6.2: Flowchart showing the steps generally followed in induced breeding through hypophysation.

Factors influencing the spawning of fish after induced breeding: Besides the above technical aspects, successful induced breeding also depends on the following environmental factors:

1) **Temperature:** In many fishes, warm temperature plays a significant role in stimulating the maturation of gonads, and thus regulating their ability to respond to pituitary stimulation. Major carps breeds within a temperature range of 24°C to 31°C
2) **Rain and water current:** A heavy rain is ideal for induced breeding. In the absence of rain artificial showers can be used. In addition, fresh rain water and flooded condition also play key role triggering the spawning of carps.
3) **Turbidity:** During induced breeding turbidity between 100ppm to 1000ppm is the most preferable range.
4) **Light:** It is known that enhanced photoperiodic regimes result in early maturation and spawning of fish.

6.2.4 Hapa used in induced breeding

A hatchery is a small sized tank where eggs are hatched under artificial conditions within 2-15 hrs. Hatcheries produce larval and juvenile fish primarily to support the aquaculture industry. Double cloth hatching hapas are commonly used in the fish industry for hatching purposes. These are small, rectangular tanks made up of coarse cloth, where fertilised eggs are kept for hatching. The hapa is generally fixed in the water with the help of bamboo poles. The hatchery hapa is a double walled tank with

- Outer wall: It is made of either thin or coarse muslin cloth.
- Inner wall: It is made of round mesh mosquito netting cloth.

The most frequently used cloth for a hatching hapa is 2×1×1 m in size for the outer one and 1.75 × 0.75 × 0.9 m for inner wall. Inside the hapa, a water depth of around 30 cm is maintained. Generally, 75,000-1, 00,000 eggs are kept in one hapa inside the inner wall for hatching.

After hatching, the hatchlings migrate into outer hapa from the inner hapa through the mosquito netting cloth. But, the egg shells, spoiled eggs and the dead eggs remain in the inner hapa. Thus, after hatching, the inner hapa containing the waste must be removed.

The hatchlings in the outer hapa are kept for a period of 38-40 after which they are transferred to nurseries. The main advantages are that the cost is very less and the hatchlings are never contaminated with egg shells, spoiled eggs and the dead eggs.

6.3 STOCK IMPROVEMENT

To grow and address future food security, aquaculture practices require stock improvement for acquiring desirable traits such as growth rate, feeding efficiency, disease resistance and market acceptability. Genetic engineering provides powerful tools for the sustainable development of aquaculture and fish stocks in particular through chromosome manipulation techniques like 1) gynogenesis 2) androgenesis and 3) polyploidy (triploidy and tetraploidy) 4) Transgenic fish and 5) interspecific or intergeneric hybridisation.

6.3.1 Gynogenesis

Development of an organism in which the embryo contains only maternal chromosomes is called gynogenesis. It involves denaturalisation or inactivation in the penetrating spermatozoa through irradiation (UV or gamma rays) and subsequent activation of eggs with the milt lacking genetic material.

Objectives of gynogenesis: The principal objectives behind gynogenesis are:

1) **Homozygous line:** Compare to conventional inbreeding a highly homozygous inbred line can be generated in much shorter time.
2) **Heterosis:** It is possible to achieve hybrid vigour through crossing gynogenic lines with the heterozygous stock.

Method for production of gynogens: In general gynogenesis involves the following basic steps:

1) Injection of spawning hormone like ovaprim (GnRH + domperidone) to matured female brood fishes stimulates ovulation.
2) Around an hour of the expected ovulation time of the female fishes, stripe male and mix its milt with Hank's solution (1:4).
3) Irradiate this preparation with UV-rays (200-250 ~W/cm^2 for 2 minutes at 28±1^{O}C) and store in refrigerator at 4-5^{0}C (Hussain et al. 1997).

4) Stripe and collect the eggs from the female brood fishes and mix with irradiated milt. After fertilisation, diploidy can be restored by thermal (heat or cold) or hydrostatic pressure shock. Though the efficacy of these shocks varies from species to species, a cold shock below 10°C and heat shock above 42°C has lethal effect (John et al. 1984; Reddy 1993; 1999). It can be done either by:
   a) **Meiotic gynogenesis:** In this method, shock treatment is given immediately after fertilisation (5 min) which prevent extrusion of the II polar body by suppressing metaphase II in the second meiotic division.
   b) **Mitotic gynogenesis:** In this method, shock treatment given around 20-30 min after fertilisation will block the cleavage I (Fig. 6.3).

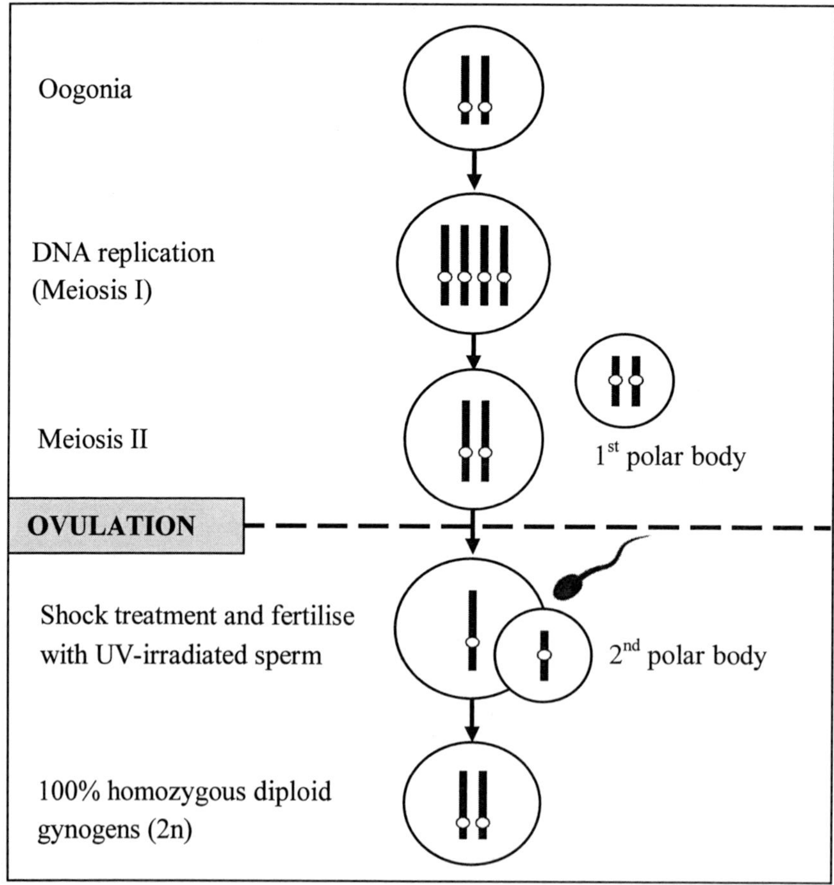

Fig. 6.3: Mechanism of inducing meiotic gynogenesis in Indian carps (IMCs).

6.3.2 Androgenesis

Development of an organism in which the embryo contains only paternal chromosomes is called androgenesis.

Approaches in androgenesis: The two commonly used approaches for androgens production are (Bhise and Khan 2002):

1) **Blocking cleavage:** In this approach, irradiated eggs (lacking DNA) is fused with normal sperms to produce haploid androgen. This is followed by shock treatment at first cleavage to prevent cell division which results in fusion of two haploid (n) daughter cells to form a diploid (2n) cell.
2) **Fusion with diploid sperm:** This approach involves fusion of irradiated eggs (lacking DNA) with sperm (2n) from a tetraploid male (4n).

Method for production of androgens: In general androgenesis involves the following basic steps (Bhise and Khan 2002):

1) **Injection of ovaprim and collection of gametes:** Inject ovaprim into mature, healthy orange males and pigmented females followed by the collection of eggs and sperms by stripping method.
2) **Inactivation of genetic material:** Irradiate eggs with UV rays by keeping eggs in an ice tray and maintaining a temperature of 4^0C to inactivate the genome of female gamete (Fig. 6.4).
3) **Fertilisation:** Fertilise irradiated eggs with a dilute solution of milt (1 (milt): 3 (physiological saline) collected from male fish. Adding about 10 ml of freshwater often stimulate fertilisation.
4) **Blocking cleavage:** Provide heat shocks to fertilised eggs using a thermo-regulated water bath to block cleavage and to induce diploidy.
5) **Hatching of androgens:** Transfer these developing embryos to aquaria filled with clean, aerated water for hatching.

Significance of androgenesis and gynogenesis: The significance of androgenesis particularly in stock improvement is briefly discussed below:

1) **Conservation of endangered species:** Using androgenesis cryopreserved sperms of extinct and endangered species can be used to fertilise irradiated eggs of closely related species, and thus germplasm of these species can be recovered.
2) **Production of monosex population:** Androgenesis can be used for species with male homogamety called monosex population. This can

be done simply by blocking the first cleavage after UV-inactivation of egg chromosomes. For example- male tilapia is known to grow faster than female. Thus, androgenesis can be used to produce monosex population of male for increasing productivity.

3) **Production of homozygous male and female:** In male heterogametic species, androgenesis followed by suppression of the first cleavage produces 50% homozygous female (XX) and 50% homozygous male (YY) offspring. These homozygous males are viable in goldfish and medaka (Yamamoto 1964; 1975)

4) **Monitoring mitochondrial inheritance:** Performance of fish with identical nuclear genotypes and different mitochondrial genotypes can be compared using androgenesis as mitochondrial DNA in animals is maternally inherited.

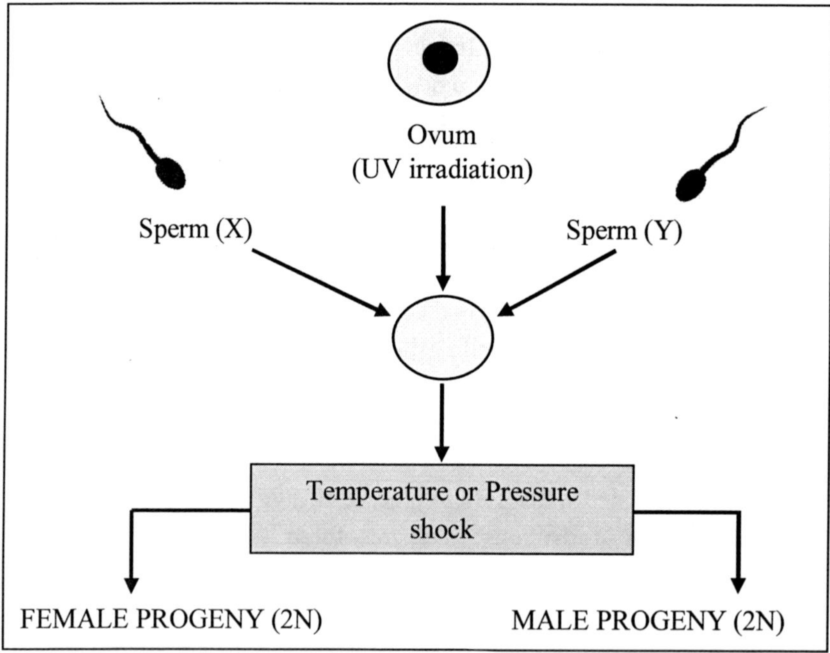

Fig 6.4: Process of development of androgenic male or female fish.

6.3.3 Polyploidy

Polyploidy is the genetic condition where an individual has one or more extra sets of chromosomes. Polyploidy becomes a promising biotechnological tool for increased production of food from aquaculture and creation of sterile organisms.

Methods of production of polyploid: Polyploidy in fish can occur either naturally or it can be induced artificially.

1) **Natural polyploidy:** There is no report of neither interspecific nor intergeneric hybrid crosses among Indian carps to produce allotriploids. However, polyploidy has been observed to occur in nature in common carp and trout mainly due to chromosomal translocation and when two distantly related fish species are crossed. For example, crosses between grass carp and bighead carp produce triploid hybrids (Marian and Krasznai 1978).
2) **Artificial polyploid:** Shock treatment by heat, cold, pressure or chemicals (colchicines, nitrous oxide, high pH, by high calcium) to eggs soon after fertilisation (within 1-5 minutes) are the common approaches in the artificial method of induction of polyploidy in fish.

   The underlying principle behind artificial polyploidy is the suppression of the separation of sister chromatids (anaphase II) by shock treatment. This prevents the formation of second polar body and hence induces the diploid condition instead of maintaining the haploid. In the subsequent step triploid zygote can be produced simply by fusing the diploid egg with haploid spermatozoa (Fig. 6.5).

Significance of polyploidy: Some of the common remarkable advantages of the polyploid condition over the diploid condition are as follows:

1) **Gigas effect and heterosis:** Polyploid fish are generally larger and they tend to live longer and grow faster than the normal diploid that is they show both gigas effect and heterosis.
2) **Controlled population:** As polyploids are generally sterile, and thus they are very useful for stocking natural water bodies of water where population control is one of the key factors.
3) **Remarkable somatic growth:** In normal diploid fishes reproductive organs are large and often accounts for more than 30% of the gross weight. In contrast to this, the energy used for normal gonadal development can be diverted to body growth in sterile polyploid.
4) **Less prone to disease:** Sexual maturation in fish is directly linked to higher infection rates, most likely due to stress. Thus, sterile triploid with underdeveloped gonads tends to have fewer diseases.
5) **Controlling exotic species:** To minimise competition with native fish, sterile triploids of exotic species could be safely used for different purposes. For example, the introduction of triploid grass carps into the reservoirs of the USA to control the growth of

unwanted macrophytes. Similarly, triploid bighead carp and silver carp were introduced into the US in sewage treatment plants and aquaculture ponds to produce clearer water.

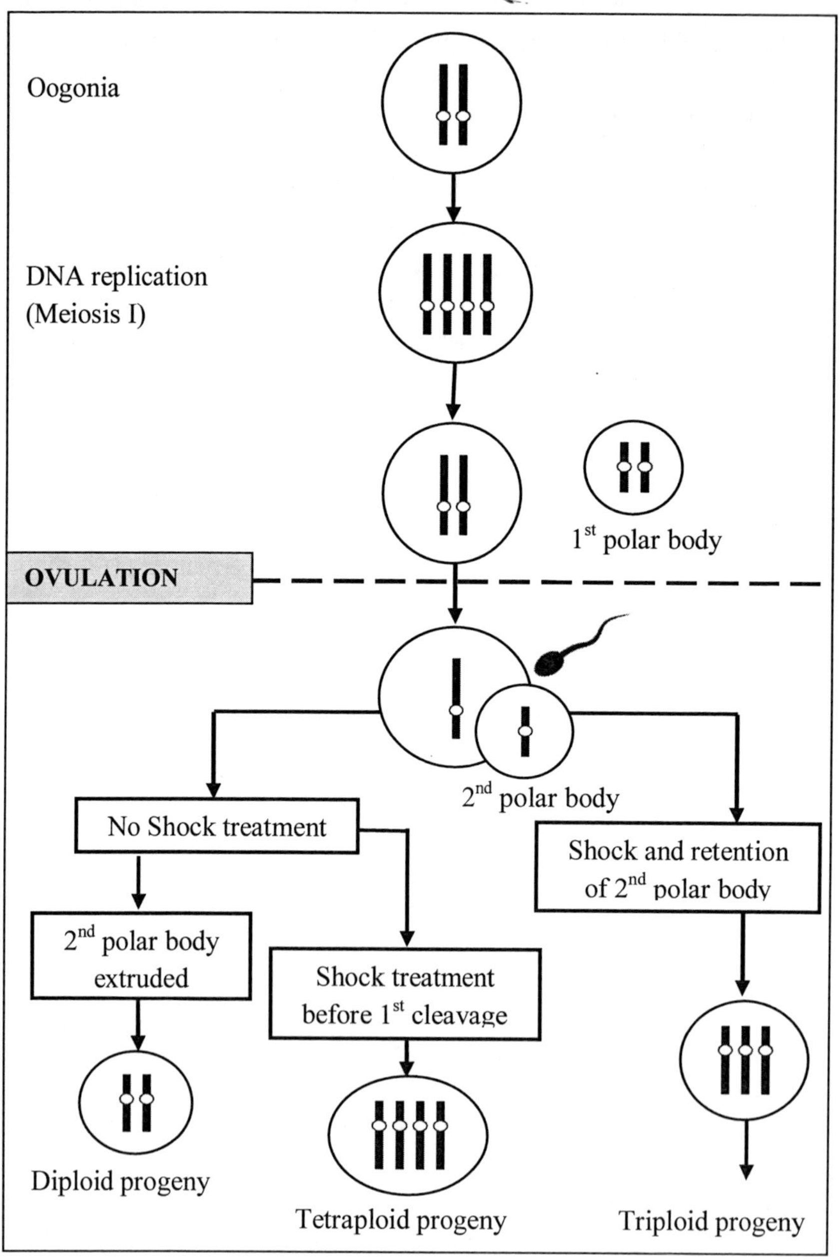

Fig. 6.5: Mechanism of inducing polyploidy (triploidy & tetraploidy) in fish in Indian carps.

6.3.4 Transgenic fish

Transgenic fish refers to the genetically modified fish in which an exogenous gene (transgene) for a desired trait has been stably integrated into its genome. The foreign gene that is inserted is called transgene and the method of production of transgenic fish called transgenesis.

Method of production of transgenic fish: The sequential steps employed in the production of transgenic fish are briefly discussed below:

1) **Choice of target gene:** The most popular gene used in aquatic species includes- growth hormone (GH) gene, antifreeze proteins (AFP) or glycoproteins (AFGP), gene for disease resistance like lysozyme, infectious hematopoietic necrovirus (IHNV) glycoprotein, human lactoferrin and cecropins (antimicrobial peptide) and firefly luciferase, etc.
2) **Cloning the gene of interest:** Clones of the gene of interest is generally generated by ligating the construct into a bacterial plasmid and then its propagation into within the bacterial cells. Using restriction enzyme clones of the gene construct are cut out from recombinant plasmid for insertion into the eggs of the host species (Fig. 6.6).

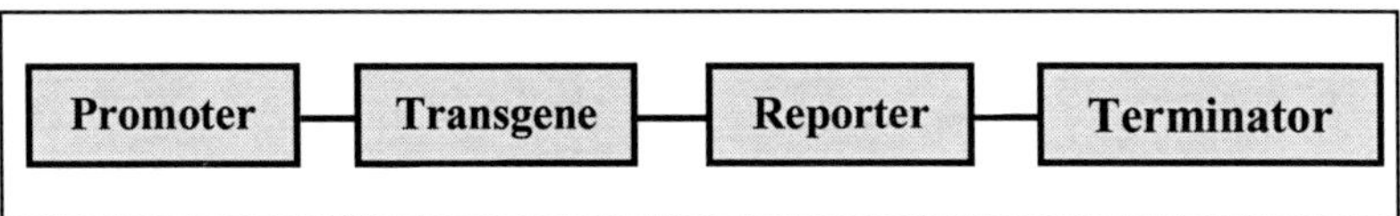

Fig. 6.6: Schematic representation of a DNA construct for a "transgene".

3) **Insertion of gene construct:** The common approaches employed in transgenesis for gene construct insertion are:
   a) **Retroviral vector method:** In this method, pantropic retroviral vector is used to deliver the small gene of interest. However, this method is not in regular use due to the risk of retroviral contamination.
   b) **Microinjection method:** It is the most common and easiest method for gene transfer. As in fish, the egg nucleus cannot be seen using light microscope due to the opaqueness of the cytoplasm; many prefer to inject the gene of interest into the cytoplasm soon after fertilisation. The only disadvantage

associated with this delivery method is that high risk of egg damage during microinjection, and thus it requires a great deal of skill (Fig. 6.7).

c) **Electroporation:** It utilises a series of short electrical pulses to make the membrane porous and permeable to the DNA present in the surrounding media. However, this method is unpopular because of the low gene transfer efficacy.

Embryonic stem cells (ESC) are also used as a target for transgene delivery in fishes. As these cells are totipotent, they can be manipulated in vitro and subsequently reintroduced into early embryos where they can contribute to the germ line of the host (Melamed et al. 2002).

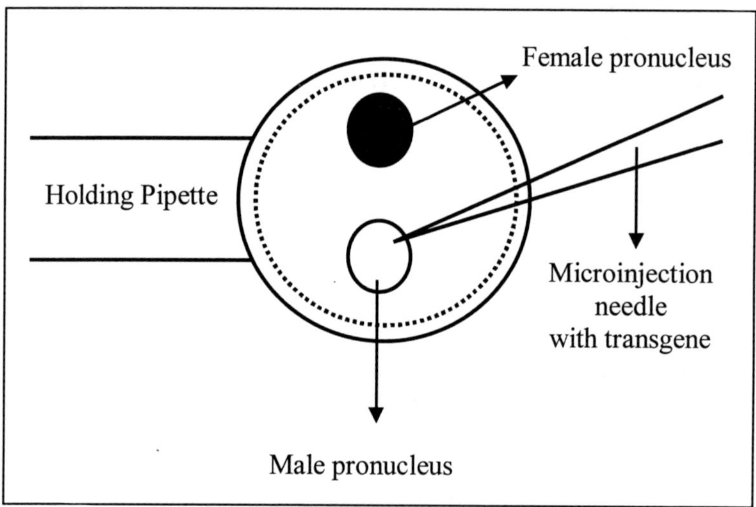

Fig. 6.7: Microinjection of gene construct into fish egg.

Application of transgenic fish: The transgenic fish has been widely used in biotechnology for research, enhancing traits of commercially available fish, production of biomedically important proteins, recreation, etc. Some of the applications of transgenic fish are briefly discussed below:

1) **Recreation:** Transgenic ornamental fish has great value in the pet market. For example, transgenic tetra, barb, zebrafish (expressing genes from jellyfish and sea coral) with bright red, green, orange-yellow, blue, and purple fluorescent colours developed through GloFish technology are the first genetically modified animal publicly available as a pet.

2) **Food:** The most successful application of the transgenic technology in aquaculture is to generate fast growing transgenic fish. In several species of salmon, trout and tilapia expression of growth hormone has resulted in dramatic growth. Beside this meat production in GM fish can be increased by double muscling. It is achieved by expressing the follistatin gene in GM fish where it inhibits myostatin which inhibits muscle cell growth and differentiation.

3) **Biosensor:** Transgenic fish are also used in environmental monitoring. Zebrafish have been used for years for toxicology and water pollution studies. Transgenic zebrafish containing firefly genes light up when exposed to carcinogenic polychlorinated biphenyls, and thus GM zebrafish can be used as a biosensor for environmental toxicity monitoring.

   Transgenic fish carrying metal-responsive or heat-shock promoters spliced into green fluorescent or LUC reporter genes provide an in vivo indicator of exposure to heavy metal pollutants (Winn 2001). Transgenic zebrafish carrying CYP-green fluorescence protein (Cytochrome P450-GFP) construct, driven by CYP1A1 promoter shows green fluorescence signals according to concentrations PCBs making it a live biosensor for polychlorinated biphenyls toxicity in the water body (Hung et al. 2012).

4) **Disease resistant fish:** It is possible to develop transgenic disease resistant fish. For example, it has been found that insertion of the gene coding for human lactoferrin and cecropins (antimicrobial peptide) in grass carp, channel catfish doubles their survival rate compared to the control fish after exposure to pathogenic bacteria and virus. Similarly, expression of rainbow trout lysozyme gene in Atlantic salmon, the latter becomes resistant to many bacterial pathogens.

5) **Cold resistant fish:** It is possible to establish an aquaculture industry in the temperate region, where water gets frozen during winter by developing cold resistant GM fish variety. For example, antifreeze protein genes from winter flounder have been introduced into Atlantic salmon to increase their cold tolerance (Shears et al. 1991).

6) **Bioreactor:** Fish muscle bioreactor system can be used for expression of medically important proteins. The advantages of the transgenic fish bioreactor include the speed of generation of transgenic fish, low cost, and low risk of transmitting animal pathogens, etc. Furthermore, as

the fish muscle is edible and raw meat may be directly consumed, it has a great potential for developing the edible vaccine (Gong and Hew, 1995).

7) **Model organism:** Zebrafish and medaka are the two common GM fish used in basic research in genetics and development. The robust regenerative capacity of zebrafish in a variety of tissues, including the fin, spinal cord, retina, and heart, making it the sole regenerative vertebrate organism which draws attention of biologists for genetic manipulation for unlocking human organ tissue diseases, failure mysteries including heart tissue repair and regeneration (Major and Poss 2007).

   As a member of vertebrate, fish basically has the same set of genes as a human does, and thus molecular mechanisms of carcinogenesis are assumed to be more or less identical between fish and human. In addition to this, easy availability of a large number of zebrafish and low cost of experiments makes zebrafish as a new cancer model in the past few years (Stern and Zon 2003).

6.3.5 Fish hybridisation

Hybridisation is a fish breeding technique. The mating or crossbreeding between animals of different breeds, varieties, species (inter-specific) or genera (inter-generic) through sexual reproduction is called hybridisation. The offspring resulting from the hybridisation is called a hybrid or cross-breed.

Hybridisation aims to combine the positive traits of the parent species into their hybrid offspring. The positive traits may be better growth rate, improve productivity through hybrid vigour, reduced unwanted reproduction through the production of sterile fish, resistance to disease or the environment, better fecundity, better food conversion capability, etc. (Rahman et al. 2013). Hybridisation can be of the following three types:

1) **Interspecific hybridisation:** Crossbreeding between different species of the same genus is called interspecific or intrageneric hybridisation.
   Example- *Labeo rohita* × *Labeo calbasu* = Rohu calbasu
2) **Intergeneric hybridisation:** Crossbreeding between individuals belongs to different genus is called intergeneric hybridisation.
   Example- *Catla catla* × *Labeo rohita* = Catla rohita

3) **Intervarietal or intraspecific hybridisation:** Crossbreeding between individuals of the same species, but belonging to different populations or breeds is called intraspecific hybridisation.

Prospects of fish hybridisation: A hybrid with selected or favoured characteristics of each parent or characteristics superior to both parents (hybrid vigor or positive heterosis) is the main goal of fish hybridisation. Compare to other vertebrate animal groups, natural hybridisation occurs widely in fishes. This high rate of natural hybridisation in fish is mainly due to external fertilisation, weak behavioural isolating mechanisms, an unequal abundance of the two parental species, competition for limited spawning habitat and poor habitat complexity (Rahman et al. 2013). The potentiality of artificial hybridisation in the world's aquaculture production is briefly described under the following headings:

1) **Improved growth rate:** The most desirable trait for stock improvement in fishery industry is the increased growth rate of hybrid. Some of the examples of successful hybridisation experiments where hybrid exhibit faster growth rate are:
    - White bass (*Morone chrysops*) × striped bass (*M. saxatilis*) (Smith and Jenkins 1984).
    - Mrigal (*Cirrhinus mrigala*) × Catla (*Catla catla*).
    - Common carp (*Cyprinus carpio*) × Rohu (*Labeo rohita*) (Khan et al. 1990).
    - Catla (*Catla catla*) and Fimbriatus (*Labeo fimbriatus*) (Basavaraju et al. 1995).
2) **Production of sterile hybrid:** In some cases production of sterile hybrid is advantageous either to diminish unwanted reproduction or to avoid energy loss due to prolific breeding. For example, grass carp is used in biological control in many areas to control the higher aquatic vegetation. But, in some areas introduction of grass carp is not permitted as its uncontrolled reproduction may eradicate aquatic plant completely. In this case, introduction of sterile hybrid can be recommended.
    - Grass carp (*Ctenopharyngodon idella*) × bighead carp (*Aristichthys nobilitrix*) = sterile hybrid (Allen Jr. and Wattendorf 1987)
    - Indian major carps (rohu, catla, mrigal) × common carp = sterile hybrid

3) **Production of mono-sex populations:** It has been found in some species that growth rate is dependent on sex. For example, male tilapia grows faster than females, female salmonids and sparids grow better than males, etc. Thus, monosex culture is one of the most promising practices for better production, while avoiding uncontrolled reproduction and overcrowding.
   - Nile tilapia (*Oreochromis niloticus*) × the blue tilapia (*O. aureus*) = male offspring.
   - Crosses between Nile tilapia (*O. niloticus*) and Wami tilapia (*O. honorum*), Nile tilapia and long-finned tilapia (*O. macrochir*), Mozambique tilapia (*O. mossambicus*) and Wami tilapia (*O. honorum*), *Tilapia mossambica* (African stock) and *T. mossambica* (Malaysian stock) produce hybrid male offspring with excellent growth rate (Wohlfarth 1994).
   - Striped bass (*Morone saxatilis*) × yellow bass (*M. mississippiensis*) = female offspring.
4) **Production of improved variety (heterosis):** It has been found that resultant hybrids often possess the superior quality to either parent or combine desirable traits from both parents with respect to size, vigour, vitality, resistance to unfavourable environmental conditions and diseases, etc. For example,
   - African catfish (*Clarias gariepinus*) × Thai catfish (*C. macrocephalus*) = hybrid with fast growth rate and desirable flesh characters (Nwadukwe 1995).
   - Catla (*Catla catla*) × Rohu (*Labeo rohita*) = rohu-catla hybrid grows almost as fast as pure catla, but has the small head of the rohu (Reddy, 2000).
   - White bass (*Morone chrysops*) × striped bass (*M. saxatilis*) = sunshine bass hybrid or whiterock bass possesses many advantageous traits like good osmoregulation, resistance to stress and disease, high thermal tolerance, high survival in culture and modified water-bodies and ability to utilise soybeans as a protein source (Smith 1988).
5) **Production of disease resistant hybrid:** It is possible to produce a disease resistant hybrid by breeding a higher resistant species with a less resistant one. For example,
   - Channel catfish × Blue catfish = hybrids are resistant to channel catfish virus disease.

- Rainbow trout (*Oncorhynchus mykiss*) × char (*Salvelinus spp.*) = hybrid is resistant to several pathogenic salmonid viruses that commonly infect salmon species (Dorson et al. 1991).
- Coho salmon (*Oncorhynchus kisutch*) × rainbow trout (*Oncorhynchus mykiss*) = hybrid shows increased disease resistance to a variety of salmonid viruses.

6) **Production of adverse environment tolerant hybrid:** Many studies suggested hybrid often achieve increased environmental tolerance, particularly when one parental species has a wide range of environmental tolerance limit. For example,
   - Mozambique × Nile tilapia = hybrid red tilapia and Mozambique × Wami tilapia = Florida red strains are salinity tolerance (Lim et al. 1993; Earnst et al. 1991).
   - Lake trout (*Salvelinus namaycush*) × brook trout (*S. fontinalis*) = hybrid is tolerant to low pH of 4.9-5.4 (Snucins 1993).

7) **Hybrid polyploidization:** Chromosome manipulation (polyploidization) of hybrid often gives increased viability, growth and survival. Triploidization of the hybrid of Atlantic salmon (*Salmo salar*) x brown trout (*S. trutta*) confers increased survival and growth rate in the hybrid. Similarly, triploid of the hybrid (Pacific salmon hybrid) of chum salmon (*Oncorhynchus keta*) and chinook salmon (*O. tshawytscha*) have earlier seawater acclimatisation times, triploid of the hybrid rainbow trout and char has improved disease resistance (Rahman et al. 2013).

6.4 GERMPLASM CRYOPRESERVATION

The present exceptionally high extinction rate of fish is one of the major concerns of global fish biodiversity. The combined and interacting influences of overexploitation; aquatic pollution; flow modification; destruction or degradation of habitat; and invasion by exotic species resulted in the declination of fish biodiversity worldwide (Goswami et al. 2016). In addition to various in situ conservation strategies, the majority of the threatened species; however, requires substantially greater effort to improve their status through various ex situ conservation strategies including bio-banking or germplasm cryopreservation.

Germplasm cryopreservation: It is a long-term storage technique to preserve sperm, eggs and embryos of threatened and endangered species without deterioration for at least several thousands of years. Thus, the germplasm bank provides a reliable source of fish genetic material for scientific and aquaculture purposes as well as opens a new frontier for aquaculture research and conservation (Tsai and Lin 2012).

The major steps in the cryopreservation process are- a) addition of cryoprotectants to cells/tissues before cooling b) cooling of the cells/tissues to a low temperature (−196°C) c) warming of the cells/tissues and d) removal the cryoprotectants from the cells/tissues after thawing (Gao and Critser 2000).

1) **Sperm cryopreservation:** Successful cryopreservation of fish sperm have been achieved for more than 200 freshwater and marine fish species, including carp, salmonids, rainbow trout, catfish, cichlids, medakas, whitefish, pike, milkfish, grouper, cod, and zebrafish. National Bureau of Fish Genetic Resources (NBFGR), Lucknow, India developed species-specific sperm cryopreservation protocols for 14 fish species. Compare to fresh sperm, a fertilisation and hatching rate of 95% using the frozen-thawed sperm has been reported for the common carp (Magyary et al. 1996). However, marine frozen-thawed spermatozoa have a higher survival and fertilisation capacity compared to freshwater species (Goswami et al. 2016).
2) **Oocyte cryopreservation:** The smaller size range, lower water content and absence of a fully developed chorion in oocyte making oocyte preservation a better way to preserve the female fertility than embryo cryopreservation. However, immature oocytes are generally alternative for the mature eggs because immature oocytes are smaller in size, while the later stage oocytes has a much lower surface area to volume ratio because of large their size. This reduces water and cryoprotectant movement rate into and out of oocytes during the cryopreservation (Tsai and Lin 2012).
3) **Embryo cryopreservation:** Embryo cryopreservation is advantageous over sperm and oocyte as the embryo preservation preserves both the paternal and maternal genomes. However, multicompartmental biological systems, high chilling sensitivity, low membrane permeability and their large size, which gives a low surface area to volume ratio of fish embryo limits successful cryopreservation of intact fish embryos (Zhang and Rawson 1995). In

addition to this, studies on the chilling sensitivity of embryos of many fishes, including brown trout, rainbow trout, carp, fathead minnows, goldfish and zebrafish demonstrated that the later stages (after 50% epiboly) were less sensitive to chilling, but as the temperature fell below zero, the chilling sensitivity increased significantly. Thus, the main obstacles to achieve successful cryopreservation of these embryos are high chilling sensitivity, their complex membrane structure and large yolk content (Tsai and Lin 2012).

4) **Blastomeres cryopreservation:** Blastomeres produced as the result of cell division are totipotent and pluripotent, and thus their cryopreservation is a promising approach to preserve the genotypes of zygotes and reconstitution of the organism. Additionally, genetic diversity of both nuclear genome and mitochondrial DNA can be maintained through the cryopreservation of blastomeres (Nilsson and Cloud 1992). Cryopreservation of blastomeres has been successful in several fish species, including zebrafish, where survival rate of 84.8% was obtained after cryopreservation of 50% epiboly zebrafish blastomeres (Harvey 1983). Blastomeres cryopreservation was also carried out in rainbow trout, carp and medaka. After post-thawing, a survival of 95% was obtained from rainbow trout blastomeres, initially cryopreserved through controlled slow freezing procedures (Calvi and Maisse 1998).

Method commonly used for cryopreservation of fish germplasm: Due to the large volume of yolk and impermeable perivitelline membranes of egg and embryo, unlike spermatozoa, cryopreservation of fish eggs and embryos is comparatively unsuccessful. A generalised method commonly followed in cryopreservation of sperm is briefly discussed below:

1) **Collection of milt:** Sperm can be collected either by stripping method or by surgical removal of testis from fish (catfishes and murrels) in which sperm cannot be stripped. In stripping method, sperm can be collected by gently massaging the abdomen of anaesthetized brood fish. In the latter method, the surgically removed testis is crushed or sliced to release sperm. Depending on species and concentration, a suitable volume of the extender is generally used to dilute the milt.
2) **Selection of extender:** Extender is a solution of which helps to maintain the viability of cells in a non-activated state during refrigeration. For example, Hanks' balanced salt solution (HBSS) is a

common extender used for catfishes. This extender can be noncryogenic for short term storage or cryogenic for long term storage. The cryogenic extender contains cryoprotectants that minimise the stress on cells during cooling and freezing.

The common permeating cryoprotectants include DMSO (Dimethylsulphoxide), methanol and glycerol and non-permeating cryoprotectants include sugars such as glucose or sucrose, polymers such as dextran, milk and egg proteins and antifreeze proteins such as those found in polar fishes.

3) **Preparation of diluents:** Cryopreservation efficacy is greatly enhanced if the prefrozen milt is diluted with a suitable extender and cryoprotectant. Diluent is prepared by mixing a suitable ratio of extender and cryoprotectant.

   For example, mixing 10 ml of glycerol, 5 ml of DMSO and 90 ml of extender. A working ratio of milt and diluent of 1:4 and 1:5 is generally used for carps and tilapia.

4) **Equilibration:** The time interval between mixing the milt with the diluents and putting the mixture into liquid nitrogen is called equilibration time. This must be sufficient enough to facilitate the penetration of cryoprotectants into cells. It depends on the type and concentration of cryoprotectant being used. For example, for rapidly permeating cryoprotectants such as DMSO, methanol the equilibration time should be in between 15 to 30 mins.

5) **Ampouling:** In this step, the diluted samples within the equilibration period is filled either in French straw or visitubes and plugged with sealing power.

6) **Freezing:** Care should be taken during freezing step. For example, it should be rapid so that thermal shock is minimal, but should not so fast to allow the formation of cellular ice crystals.

   For example, the sealed visitubes are generally cooled at 15°C/minute by a programmable cooling chamber.

7) **Storing:** The visitubes are stored in liquid nitrogen at -196^{0}C and. During storing period, the evaporation loss of liquid nitrogen should be regularly compensated by regular re-filing.

8) **Thawing:** The success of cryopreservation also depends on cooling and thawing rate. For example, fast thawing (38^{0}C for 7-9 seconds) is preferable for carp spermatozoa as slow thawing can re-crystalise the small intracellular crystals which may damage the cells.

Risks associated with cryopreservation: The freezing stage is the most crucial stage in the process of cryopreservation due to higher chances of cell damages. The damages mainly occur due to solution effects, extracellular ice formation, dehydration and intracellular ice formation; however, these problems can be minimised by using suitable concentration of cryoprotectants like DMSO, glycerol, methanol etc.

1) **Solution effects:** High concentrations of solutes may result in loss of stability in the membranes, denaturation of proteins, solute loading.
2) **Extracellular ice formation (EIF):** Slow-freezing results in the efflux of water from the cells and formation of ice in extracellular space which can cause mechanical damage by crushing of cell membranes.
3) **Dehydration:** Migration of water in extracellular space also causes cellular dehydration results in cell damage.
4) **Intracellular ice formation (IIF):** Compare to extracellular ice, intracellular ice is almost always fatal to cells. Therefore, during freezing, efforts are to be taken to minimise the probability of intracellular ice forming either by cooling very rapidly or by cooling slowly (Acker et al. 2001).
5) **Chilling injury and cold shock:** Both rapid cooling (cold shock) and low temperature per second (chilling injury) can cause cellular damage due to disruption of cellular lipids and proteins. For example, low temperature can causes damage to the meiotic spindle, actin filaments, and chromosomal dispersal and microtubule depolymerisation (Prentice et al. 2011; Tamura et al. 2013).
6) **Influence of cryoprotectants:** The concentration of the cryoprotectant and the relative permeability of the membrane to water and to the cryoprotectants determine the degree to which cells shrink and re-swell after addition of a membrane-permeable cryoprotectant (Kleinhans 1998). During thawing, damage due to over-swelling of cells can be minimised by stepwise removal of the cryoprotectant.

6.5 FISH DNA BARCODING

6.5.1 DNA barcoding

DNA barcoding is a DNA-based taxonomic method. It uses short, highly variable genetic marker sequence from a standard part of the genome like cytochrome oxidase I (COI), 16S rRNA, 18S rRNA genes, ITS regions to identify biological species.

The most commonly used barcode region for almost all animal groups is the 648 base-pair region located in the mitochondrial cytochrome c oxidase 1 gene (CO1). It widely used for identifying birds, butterflies, fish, flies and many other animal groups. The main advantage of using COI is that it is short enough to be sequenced quickly and cheaply, while long or varied enough to identify variations among species. COI has advantages over nuclear genes as mitochondrial genes lack introns, are generally haploid, and exhibit limited recombination. The COI barcode region is now recognised as the official barcode marker for animals by iBOL (the International Barcode of Life Project). In the case of plants, regions of chloroplast DNA (e.g., rbcL, trnL-F, matK, trnH, psbK, etc.) are the common barcode locus.

The concept of DNA barcoding was first proposed by Paul Hebert and his colleagues in a paper titled "Biological identifications through DNA barcodes". They proposed DNA barcoding a new system of species identification and discovery using a short section of DNA from a standardised region of the genome (Hebert et al. 2003). This barcode sequence can be used to identify different species just like the way a supermarket scanner identifies the products using the black stripes of the Universal Product Code (UPC) barcode present on the products.

6.5.2 Technique of DNA barcoding

DNA barcoding, a new DNA-based method for the quick identification of biological species is comprised of three main steps: DNA extraction, PCR amplification, and DNA sequencing and analysis.

1) **Specimen:** It starts with the specimen in question. Natural history museums, herbaria, zoos, aquaria, frozen tissue collections, seed banks, type culture collections and other repositories of biological materials or natural habitats are common sources of the specimen for DNA barcoding (Fig. 6.8).
2) **Extraction of genomic DNA and amplification of barcode region:** High quality DNA extraction is a key step because PCR amplification totally depends on the purity of DNA. Genomic DNA can be extracted and purified from tissue, blood or any other processed biological material. Universal primer pairs are then used to amplify

the known 648 base-pair region of the cytochrome oxidase I (COI) gene.

3) **Sequencing:** Following the PCR amplification, the amplicon is sequenced using any of the high throughputs sequencing technology. The sequence is represented by a series of letters ATGC.

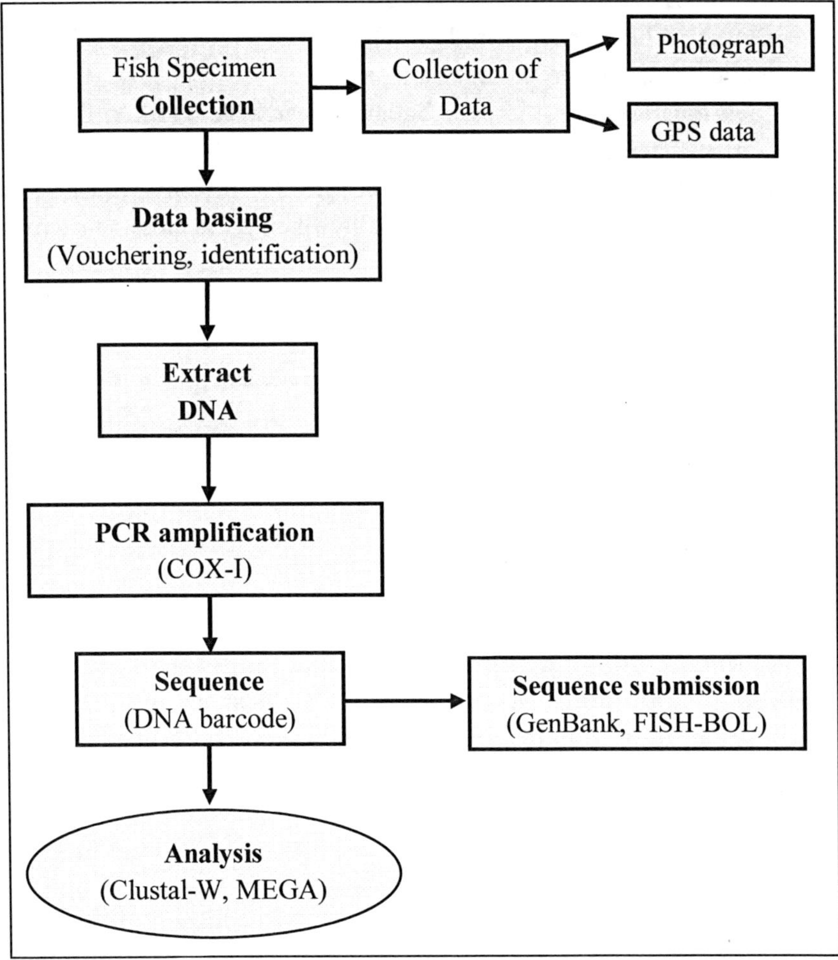

Fig 6.8: Schematic representation of the steps followed in DNA barcoding in molecular taxonomy.

4) **DNA barcode library:** One of the most important components of the barcode project is the construction of a public reference library. For example, Barcode of Life Database (BOLD, www.boldsystems.org/) was created and maintained by the University of Guelph in Ontario. It

offers researchers a way to collect, manage, and analyse DNA barcode data.

The Consortium for the Barcode of Life (CBOL) is an international initiative of 200 organisations from more than 50 countries, which aim to support the development of DNA barcoding as a global standard for species identification. As of November, 2016, BOLD included over 5 million DNA barcode sequences, of which around 174,327 were from the animal.

5) **Identification:** Basic Local Alignment Search Tool (BLAST) of the National Center for Biotechnology Information (NCBI) and Barcode of Life Data Systems (BOLD) is then used to identify sequences in databases. Specimens are identified by finding the closest matching reference record in the database.

   Further, multiple sequence alignment (e.g., ClustalW, ClustalX, ClustalΩ, etc.) and tree-building tools (e.g., MEGA, PHYLIP, CVTree, etc.) can be used to analyse phylogenetic relationships (Fig. 6.8).

6.5.3 DNA barcoding in fish identification

The benefits of fish barcoding include quick identification of fish for all potential users, processed specimens, identification of previously unrecognised species; and identifications of species like cryptic or sibling species, where traditional methods are incapable of identifying those (Ward et al. 2009).

FISH-BOL, the Fish Barcode of Life, is a global research collaboration initiated in 2005 which aims to develop a reference barcode sequence library for all fish species. In other words, FISH-BOL (http://www.fishbol.org/) initiative simply seeks to fish identification using the 648 base pair barcode region of the mitochondrial COI gene.

As of 2016, FISH-BOL contains DNA barcodes for more than 10000 species, images, and geospatial coordinates of examined specimens. The database also contains external linkages to voucher specimens, information on species distributions, nomenclature, authoritative taxonomic information, collateral natural history information and literature citations (Becker et al. 2011).

In India, Fish Barcode Information System (FBIS) has been developed by National Bureau of Fish Genetic Resources (NBFGR), Lucknow for fishes of the Indian subcontinent. FBIS contains COI barcode records of fishes collected from various water bodies of India. Thus, FBIS complements the existing DNA barcode sequence database, like FISH-BOL. As of November, 2016 FBIS contains specimen records of 24142 of 1595 species belonging 220 families (Nagpure et al. 2012).

6.5.4 Advantages and disadvantages of DNA barcoding

DNA barcoding method for species identification is based on partial sequences of the mitochondrial cytochrome c oxidase 1 gene (COI). The method offers several advantages over conventional identification method like:

1) Fishes are the most diverse group of vertebrates and there may be more than 30,000 different species. However, correct fish identification is no easy task because of their high diversity and profound changes in appearance during various stages of their development. In this context, DNA barcoding which is based on partial sequences of the mitochondrial COI gene has its applicability for all life stages.
2) The conventional identification method which is primarily based on morphology fails in case of the fragmentary or processed remains. DNA barcoding, on the other hand, can be used for identification and traceability of seafood, meat, edible plants, dairy products and processed foods.
3) DNA barcoding is able to differentiate phenotypically alike species. For example, DNA Barcoding is used to reveal the existence of cryptic species in the neotropical skipper butterfly (ten species in one), Cuban freshwater fishes, Pencilfishes (Benzaquem et al. 2015).
4) The method is also advantageous as it requires a very small amount of biological samples without sacrificing the animal.
5) The DNA Barcoding approach can be used for identifying ornamental fish trade. This will help in monitoring the trafficking of threatened species by fish poachers.

On the other hand, it is difficult to resolve recently diverged species or new hybridised species through DNA barcoding. Additionally, there is no universal gene for DNA barcoding, no single gene that is conserved in all domains of life and exhibits enough sequence divergence for species discrimination. Thus, the validity of DNA barcoding completely depends on initial establishment of reference sequences from taxonomically confirmed specimens (Purty and Chatterjee 2016).

6.6 BIOTECHNOLOGY IN AQUACULTURE: ITS PROSPECTS AND CHALLENGES

Demand for fish is soaring globally, and thus pisciculture becomes the world's fastest-growing sector of agriculture. At the same time, wild fish stocks are declining rapidly, mainly because of overfishing and fish habitat destruction due to destructive fishing techniques like bottom trawling, dynamiting, and poisoning.

Aquaculture animals are particularly well suited for research in biotechnology mainly due to the availability of large numbers of germ cells, external fertilisation system, easy hormone induced sterility or functional sex reversal, etc. Thus, biotechnology can make a great contribution to improve aquaculture yields through sustainable development and to meet the present crisis of animal protein from aquatic habitat. Some of the major contributions of biotechnology in direction of sustainable development of fish farming are discussed below (Lakra and Ayyappan 2003; Shankar Murthy and Kiran 2013)-

1) **Transgenic fish (GMOs):** The transgenic fish has been widely used in biotechnology for enhancing traits of commercially available fish (disease, cold resistant fish), production of biomedically important proteins, recreation, etc. For example, in several species of salmon, trout and tilapia expression of growth hormone gene has resulted in dramatic growth. Similarly, expression of antifreeze protein gene from winter flounder into Atlantic salmon makes them cold resistant (Fig. 9).
2) **Fish breeding:** Successful breeding of economically important fishes which generally do not breed in captive condition makes a revolutionary in fish farming. Breeding through ovaprim (sGnRHa/ OvaRH + domperidone/ pimozide) injection is now the best available biotechnological tool for the induced breeding of fish.

Ovaprim is a stable solution that contains OvaRH or sGnRHa (a synthetic analogue of Salmon GnRH) and a dopamine inhibitor like domperidone. Ovaprim utilizes the fish's own endocrine system to safely induce maturation and coordinate spawning dates. The benefits of ovaprim injection are-

- Induce maturation within a spawning season.
- Advance spawning date.
- Coordinate and synchronize spawning times when used in the normal spawning cycle.
- Increase milt production including increased sperm count.
- Induce maturation in difficult species.
- Induce spawning in the most important and difficult to spawn species.

3) **Fish hybridisation:** It is possible to combine the positive traits of the parent species into their hybrid offspring through hybridisation. E.g., hybrids of white bass and striped bass, mrigal and catla have increased growth rate.
4) **Chromosome engineering:** Polyploidy (triploid, tetraploid) becomes a promising biotechnological tool for increased production of food from aquaculture and creation of sterile organisms. The culture of triploid fish can be advantageous for several reasons like increased growth rate, poor reproductive growth, while increased somatic growth (broiler fish), controlled population size due to their sterility, diseases resistance, etc.
5) **Probiotics and bio-remediation:** There is growing interest in the use probiotics and bioaugmentation technology in aquaculture. Bioaugmentation is a new bioremediation strategy in which useful microbes (*Nitrosomonas*, *Nitrobacter*, etc.) are inoculated to improve pond bottom by converting the ammonia into nitrate. The probiotic technique allows the growth of bacteria beneficial for host or habitat, while suppressing the other.

   Probiotics improve productivity by inhibiting the growth of pathogenic bacteria, improving water quality, improving nutrient digestion (secreting enzymes), improving stress tolerance, etc. Some of common probiotics used in aquaculture technique are- *Lactobacillus acidophilus*, *Bacillus subtilis*, *Bacillus licheniformis*, *Debaryomyces hansenii*, *Bacillus cereus*, *Amphiprion percula*, etc.

6) **Uniparental chromosome inheritance:** Both androgenesis and gynogenesis have particular interest to fish breeders. These techniques can be used for recovery of lost genotypes from the cryopreserved gametes, generation of homozygous lines and generation of a monosex population where a particular sex has beneficial a trait like increased growth rate.

7) **Monosex culture:** Sex control technique becomes an important biotechnological tool to increase aquaculture production. A monosex population can be obtained by exposing juveniles to a temperature of 37^0C (Nile tilapia) or 35^0C (Blue tilapia) over 28 days after yolk sac resorption, sex steroids injection during the early phase before sex differentiation starts or by hybridisation. It improves productivity, particularly where one sex is more useful than the other. For example, male tilapia grows faster than females, female salmonids and sparids grow better than males, etc.

8) **Cryopreservation and fish cell line:** The present exceptionally high extinction rate of fish is one of the major concerns of global fish biodiversity. Cryopreservation overcomes problems of male maturing before the female, allow selective breeding, stock improvement and enables the ex-situ conservation of rare and endangered species.

   Fish cell lines (e.g., RTCi-2, CHSE-214, FHM, BB, BF-2, etc.) derived from carp, catfish, salmonids, mullets, etc. are widely used in aquaculture for study and diagnosis of fish viruses, production of viral vaccines and aquatic toxicology. Fish cell lines are also used for studying various cellular and physiological processes like studying the production of growth hormone prolactin in the cell line derived from the pituitary gland of eel, tilapia and rainbow trout.

9) **Sustainable aquaculture:** Farming of new generation fish not only increases the productivity, but also protects water quality, marine sediments and aquatic health. For example, half the time required to raise salmon compared to the wild means supply can be increased without proportionately increasing the use of coastal waters, increase in FCR means that fewer natural resources are required to produce the fish, etc.

10) **Fish health management:** Biotechnological tools such as molecular diagnostic methods, use of vaccines and

immunostimulants are now very popular for their significant role in improving the disease resistance in fish.

- Sensitive and specific tools like gene probes and PCR-based diagnostic methods are widely used for disease diagnosis.
- A number of vaccines against bacteria and viruses (e.g., furunculosis) have been developed for finfish fisheries.
- Injection of immunostimulants like glucan, chitin and levamisole stimulates phagocytic activities and specific antibody responses in fish.
- Monoclonal antibodies (MABs) developed (hybridoma technology) against several viral and bacterial pathogens of fish can be used for rapid diagnosis endemic diseases through ELISA technique.

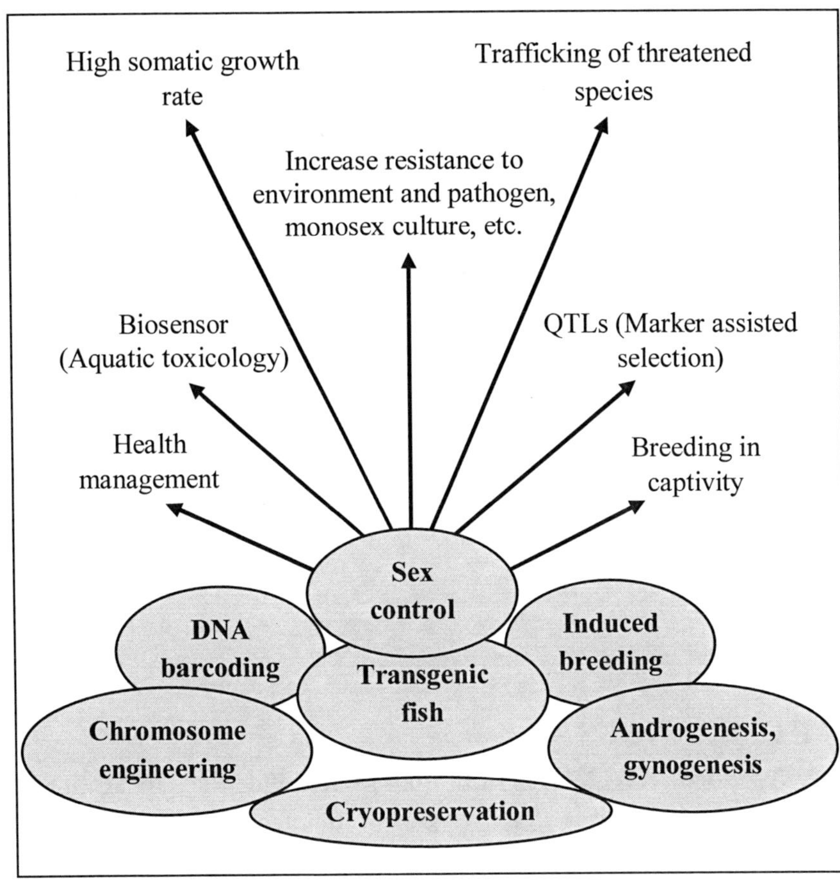

Fig. 6.9: Prospects of aquaculture biotechnology

11) **Molecular marker:** DNA markers have a significant impact on the genetic improvement of aquaculture species. For example,

- The 648 base-pair region of the mitochondrial cytochrome oxidase I (COI) gene (DNA barcoding) can be used for identifying fish trade. This will help in monitoring the trafficking of threatened species by fish poachers.
- Multi-locus VNTR analysis (DNA fingerprinting) can be used to assess the amount of inbreeding in cultured populations.
- Genetic markers can be used to identify individuals and family groups so that they can be reared together thus simplifying experimental designs.
- Quantitative trait loci or QTLs can be used for marker-assisted selection (MAS) in fish.

12) **Environmental monitoring:** Transgenic fish are also used in environmental monitoring (biosensor). Zebrafish have been used for years for toxicology and water pollution studies.

Transgenic zebrafish containing firefly genes light up when exposed to carcinogenic polychlorinated biphenyls, and thus GM zebrafish can be used as a biosensor for environmental toxicity monitoring. Transgenic fish carrying metal-responsive or heat-shock promoters spliced into green fluorescent or LUC reporter genes provide an in vivo indicator of exposure to heavy metal pollutants.

6.7 REFERENCES

Acker JP, Elliott JAW, McGann LE (2001) Intercellular ice propagation: experimental evidence for ice growth through membrane pores. Biophysical Journal. 81(3):1389–1397.

Akankali JA, Seiyaboh EI, Abowei JFN (2011) Fish breeding in Nigeria. Advance Journal of Food Science and Technology 3(2): 144-154.

Allen SKJr and Wattendorf RJ (1987) Triploid grass carp: Status and management implications. Fisheries 12(4): 20-24.

Baroiller JF and D'Cotta H (2001) Environment and sex determination in farmed fish. Comparative Biochemistry and Physiology. 130(4): 399-409.

BasavarajuY, Devaraj KV, Ayyar SP (1995) Comparative growth of two carp hybrids between *Catla catla* and *Labeo fimbriatus*. Aquaculture 129(1):187-191.

Becker S, Hanner R, Steinke D (2011) "Five years of FISH-BOL: brief status report". Mitochondrial DNA. 22(Suppl 1): 3–9.

Benzaquem DC, Oliveira C, Batista Jda S, Zuanon J, Porto JI (2015) DNA barcoding in pencilfishes (Lebiasinidae: Nannostomus) reveals cryptic diversity across the Brazilian Amazon. PLoS One. 10(2): e0112217.

Bhise MP, Khan TA (2002) Androgenesis: the best tool for manipulation of fish genomes. Turkish Journal of Zoology. 26: 317-325.

Calvi SL, Maisse G (1998) Cryopreservation of rainbow trout (*Oncorhynchus mykiss*) blastomeres: Influence of embryo stage on post-thaw survival rate. Cryobiology. 36(4): 255-62.

Dorson M, Chevassus B, Torhy C (1991) Comparative susceptibility of three species of char and rainbow trout x char triploid hybrids to several pathogenic salmonid viruses. Diseases of Aquatic Organisms. 11(3): 217–224.

Ernst DH, Watanabe WO, Ellington LJ, Wicklund RI, Olla BL (1991) Commercial-scale production of Florida red tilapia seed in low-and brackish-salinity tanks. Journal of World Aquaculture Society. 22(1): 36–44.

Gao D and Critser JK (2000) Mechanisms of cryoinjury in living cells. Institute of Laboratory Animal Resources Journal. 41(4): 187-96.

Gong Z and Hew CL (1995) Transgenic fish in aquaculture and developmental biology. In: Pederson RA and Schatten GP eds. Current Topics in Developmental Biology, Academic Press, San Diego, CA, pp. 175-214.

Goswami M, Mishra A, Ninawe NS, Trudeau VL, Lakra WS (2016) Bio-banking: an emerging approach for conservation of fish germplasm. Poultry, Fisheries & Wildlife Sciences. 4(1): 143.

Hebert PDN, Penton EH, Burns JM, Janzen DH, Hallwachs W (2004) Ten species in one: DNA barcoding reveals cryptic species in the neotropical skipper butterfly *Astraptes fulgerator*. Proceedings of the National Academy of Sciences of the USA. 101(41): 14812-14817.

Hung KW, Suen MF, Chen YF, Cai HB, Mo ZX, Yung KK (2012) Detection of water toxicity using cytochrome P450 transgenic zebrafish as live biosensor: for polychlorinated biphenyls toxicity. Biosensors and Bioelectronics. 31(1): 548–553.

Hussain MG, Mahata SC, Rahman MS (1997) Induction of mitotic and meiotic gynogenesis and production of genetic clones in rohu, *Labeo rohita* Ham. Bangladesh Journal of Fisheries Researchis.1(2): 01-07

John G, Reddy PVGK, Gupta SD (1984) Artificial gynogenesis in two Indian major carps, *Labeo rohita* (Ham.) and *Catla catla* (Ham.). Aquaculture. 42(2): 161–168.

Khan HA, Gupta SD, Reddy PVGK, Tantia MS, Kowtal GV (1990) Production of sterile intergeneric hybrids and their utility in aquaculture and stocking. In: Keshavanath, P. and Radhakrishnan KV eds. Carp Seed Production Technology. Special Publication of the AFS No 2. Asian Fisheries Society, Mangalor, India, pp. 41-48.

Kleinhans FW (1998) Membrane permeability modeling: Kedem–Katchalsky vs a two-parameter formalism. Cryobiology. 37(4): 271–289.

Kumar D (1992) Fish culture in undrainable ponds. A manual for extension. FAO Fisheries Technical Paper No. 325, FAO, Rome, Italy, 239 p.

Lakra WS and Ayyappan S (2003) Recent advances in biotechnology applications to aquaculture. Asian-Australasian Journal of Animal Sciences. 16(3): 455-462.

Lim C, Leamaster B, Brock JA (1993) Riboflavin requirement of fingerling red hybrid tilapia grown in seawater. Journal of World Aquaculture Society. 24(4): 451–458.

Magyary I, Urbanyi B, Horvath L (1996) Cryopreservation of common carp (*Cyprinus carpio* L) sperm II Optimal conditions for fertilization. Journal of Applied Ichthyology. 12(2): 117-119.

Major RJ, Poss KD (2007) Zebrafish heart regeneration as a model for cardiac tissue repair. Drug Discovery Today: Disease Models. 4(4): 219-225.

Marian T and Krasznai Z (1978) Kariological investigation of *Ctenopharyngodon idella* and *Hypophthalmichthys nobilis* and their cross-breeding. Aquaculture Hungarica. 1: 44-50.

Melamed PG, Zhiyuan G, Fletcher G, Hew CL (2002) The potential impact of modern biotechnology on fish aquaculture. Aquaculture 204(3-4): 255-269.

Nagpure NS, Rashid I, Pathak AK, Singh M, Singh SP, Sarkar UK (2012) FBIS: A regional DNA barcode archival & analysis system for Indian fishes. Bioinformation. 8(10): 483-8.

Nilsson EE, Cloud JG (1992) Rainbow trout chimeras produced by injection of blastomeres into recipient blastulae. Proceedings of the National Academy of Sciences of the USA. 89(20): 9425-9428.

Nwadukwe FO (1995) Hatchery propagation of five hybrid groups by artificial hybridization of *Clarias gariepinus* and *Heterobranchus longifilis* (Clariidae) using dry powdered carp pituitary hormone. Journal of Aquaculture in the Tropics. 10(1): 1-11.

Piferrera F, Beaumontb A, Falguierec J, Flajshansd M, Haffraye P, Colombo L (2009) Polyploid fish and shellfish: Production, biology and applications to aquaculture for performance improvement and genetic containment. Aquaculture. 293 (3–4): 125–156.

Purty RS and Chatterjee S (2016) DNA Barcoding: An effective technique in molecular taxonomy. Austin Journal of Biotechnology & Bioengineeringis. 3(1): 1059.

Rahman MA, Arshad A, Marimuthu K, Ara R, Amin SMN (2013) Inter-specific hybridization and its potential for aquaculture of fin fishes. Asian Journal of Animal and Veterinary Advances. 8(2): 139-153.

Reddy PVGK (1999) Genetic resources of Indian major carps. FAO Fisheries Technical Paper No. 387, FAO, Rome, Italy, 76p.

Reddy PVGK (2000) Genetic Resources of Indian Major Carps. FAO Fisheries Technical Paper No. 387, FAO, Rome, Italy,76 p.

Reddy PVGK, Jana RK, Tripathi SD (1993) Induction of mitotic gynogenesis in rohu (Labeo rohita Hamil.) through endomitosis. The Nucleus. 36(3): 106–109.

Shankar Murthy K and Kiran BR (2013) Role of biotechnology in fisheries and aquaculture: an overview. International Journal of Current Trends in Research. 2 (1): 74-86.

Shears MA, Fletcher GL, Hew GL, Gauthier S, Davies PL (1991) Transfer, expression and stable inheritance of antifreezeprotein genes in Atlantic salmon (*Salmo salar*). Molecular Marine Biology and Biotechnology. 1(1): 58-63.

Smith TIJ (1988) Aquaculture of striped bass and its hybrids in North America. Aquaculture Magazine. 14, 40–49.

Smith TIJ and Jenkins W (1984) Controlled spawning of F1 hybrid striped bass (*Morone saxatilis* X *Morone chrysops*) and rearing of F1 progeny. Journal of the World Mariculture Society. 15(1-4): 147-161.

Snucins EJ (1993) Relative survival of hatchery-reared lake trout, brook trout and F1 splake stocked in low-pH lakes. North American Journal of Fisheries Management. 12(3): 460–464.

Stern HM, Zon LI (2003) Cancer genetics and drug discovery in the zebrafish. Nature Reviews Cancer. 3(7): 533-539.

Tamura AN, Huang TT, Marikawa Y (2013) Impact of vitrification on the meiotic spindle and components of the microtubule-organizing center in mouse mature oocytes. Biology of Reproduction. 89(5): 112.

Tave D (1999) Inbreeding and brood stock management. Fisheries Technical Paper No. 392, FAO, Rome, Italy, 122p.

Tsai S, Lin C (2012) Advantages and Applications of Cryopreservation in Fisheries Science. Brazilian archives of biology and technology 55(3): 425-433.

Ward RD, Hanner R, Hebert PD (2009) The campaign to DNA barcode all fishes, FISH-BOL. Journal Fish Biology. 74(2): 29-56.

Winn RN (2001) Trangenic fish as models in environmental toxicology. Institute for Laboratory Animal Research Journal. 42(4): 322-329.

Wohlfarth GW (1994) The unexploited potential of tilapia hybrids in aquaculture. Aquaculture Research. 25(8): 781–788.

Wright JRJr, Yang H (1997) Tilapia Brockmann Bodies: an inexpensive, simple model for discordant islet xenotransplantation. Annals of Transplantation. 2(3): 72-75.

Yamamoto T (1964) The problem of viability of the YY zygotes in the medaka, *Oryzias latipes*. Genetics. 50(1): 45 – 58.

Yamamoto TA (1975) YY male gold fish from mating estrone induced XY female and normal male. Journal of Heredity. 66(1): 2-4.

Zhang T, Rawson DM (1995). Studies on chilling sensitivity of zebrafish (*Brachydanio rerio*) embryos. Cryobiology. 32(3): 239-246.

*** ***** ***

INDEX

A

B

C

D

E

F

L

M

N

O

P

R

S

T

V

W

*** ***** ***